# Virtualizing 5G and Beyond 5G Mobile Networks

For a complete listing of titles in the
*Artech House Mobile Communications Library,*
turn to the back of this book.

# Virtualizing 5G and Beyond 5G Mobile Networks

Larry J. Horner

Kurt Tutschku

Andrea Fumagalli

ShunmugaPriya Ramanathan

ARTECH HOUSE

BOSTON | LONDON

artechhouse.com

**Library of Congress Cataloging-in-Publication Data**
A catalog record for this book is available from the U.S. Library of Congress.

**British Library Cataloguing in Publication Data**
A catalogue record for this book is available from the British Library.

**Cover design by Andy Meaden Creative**

ISBN 13: 978-1-63081-930-9

Cover image courtesy of Adobe Stock/Blue Planet Studio


**685 Canton Street**
**Norwood, MA 02062**



10 9 8 7 6 5 4 3 2 1

*To Maggie, Tiago, and Eli, the future is what you make it, all the best*
*—LJH*

*To Beate, Lorenz, and Mathis, your love carries me*
*—KTT*

*To Daura, Lorella, and Tommaso, the lights of my life*
*—AF*

*To Nanda, Naren, and Shreeya, the loves of my life*
*—SR*

# Contents

# Acknowledgments

No effort of this magnitude is accomplished alone, nor can all those who contributed be properly thanked to the extent they deserve, yet some individuals merit special recognition. Apologizes in advance for any oversight that have been made in omission. First, to my coauthors, Andrea, Kurt, and Priya, thanks, we made it, and hopefully have created a friendship that will remain well into the future. We must also thank the reviewers, editors, and members of the staff at Artech House: Natalie, Isabel, and some whom remain unknown to us. Thanks for granting us the opportunity to put some of our work down into words; without you this would simply remain scattered thoughts in papers, slide decks, and side conversations over the course of our careers. I must also thank my family for their encouragement and support during the creation of this work. Finally, I need to thank by name Elizabeth A. Q. for her efforts, encouragement, and design work, your name is equal to those found on the title page in my view.

Kurt's work on this book was funded partly by the Knowledge Foundation, Sweden, through the Human-Centered Intelligent Realities (HINTS) Profile Project (contract 20220068).

*Larry J. Horner*

# Part I
# Fundamentals of Virtualization in Communication Service Provider Networks

# 1

# Virtualizing of the 5G Radio Access and Core Network

## 1.1 Introduction to Virtualizing the Mobile Network

Network virtualization has evolved as the preferred method for realizing the efficient and future-proof design, engineering, and operation of communication networks. To those new to this topic there is significant expansion needed in this opening statement, which will be provided shortly. For those already well versed in this topic we hope to provide additional insights into both the business and technical aspects of the current and future work guiding the ongoing transformation underway in this field.

### 1.1.1 The Beginning of Network Function Virtualization

Before getting too deep, a short review of how the term and vision of network virtualization came to being.

A team of authors from 13 companies [1] published a white paper titled "Network Function Virtualization, an Introduction, Benefits, Enablers, Challenges & Call for Action." This relatively short paper, numbering only 16 pages, was first published October 22–24, 2012, at the SDN and OpenFlow World Congress held in Darmstadt, Germany. Those not familiar with this white paper and interested in this topic would be well served to have a quick read of this foundational paper. The authors gifted the industry with both a framework to transform the way networks are built and operated and new terminology in which to discuss the nature of the transformation. Prior to this white paper, the terms network function virtualization (NFV) and virtualized network functions (VNFs) had not been coined or used in the context of transforming the communication network. A note of caution: new members of this community may overload or mix NFV and VNF, they are not the same but are related to each other. NFV represents a concept where a network function has been disaggregated (virtualized) from the underlying infrastructure, whereas a VNF is a specific realization of a network function in a virtualized environment. The authors of the 2012 paper paid special attention to position the NFV concept alongside another technology—software-defined networking (SDN)—that had recently been gaining favor. The concept of SDN today is often overshadowed in the context of NFV discussions, which may be viewed as simply a consolidation of the two concepts into one, although technically they remain separate concepts. SDN is the separation of a traditional control function from the data plane function on an appliance. For example, with switches or routers in the classical model the control function runs on the same systems where the links are terminated. With SDN, new routing or switching decisions are managed by a centralized controller, and the routing or switching tables in the router or switch are updated via an application programming interface (API) between these nodes. The IEEE currently maintains the term NFV-SDN in their conference proceedings [2]. Significant work has been underway by leading network operators and by the vendors supplying equipment into the network in the ensuing years to implement the vision laid out in the 2012 NFV paper. The original authors were also very sensitive to the need to include the existing ecosystem solution providers, the telecommunication equipment manufacturers (TEMs), also called network equipment providers (NEPs), in their overall strategy to transform the industry. As one should fully expect in any field of technology, there has been some evolution and expansion in the procedures employed over time. This is the case with NFV in building the interesting workloads that realize the network. (There are a large number, sometimes well over a hundred processes, running on a modern system in the network. Many are important and critical but benign, placing little or no load on the system, and a smaller number perform the majority of the functional processing that realized the core intent of the service. The latter are the "interesting workload.") Nevertheless, the underlying concept laid out in the paper has proven transformative to the network and to the vendors delivering products into that network.

Figure 1.1 is from that now-famous 2012 white paper. It shows the conceptual separation of the various layers of the NFV infrastructure (NFVi) (often called a middleware layer of software), virtualization, compute, storage and networking, and the element management system (EMS), which is a nod to the TEMs, on the left, and the management and network orchestration (MANO) on the right. The EMS is one area where time has challenged the original view;

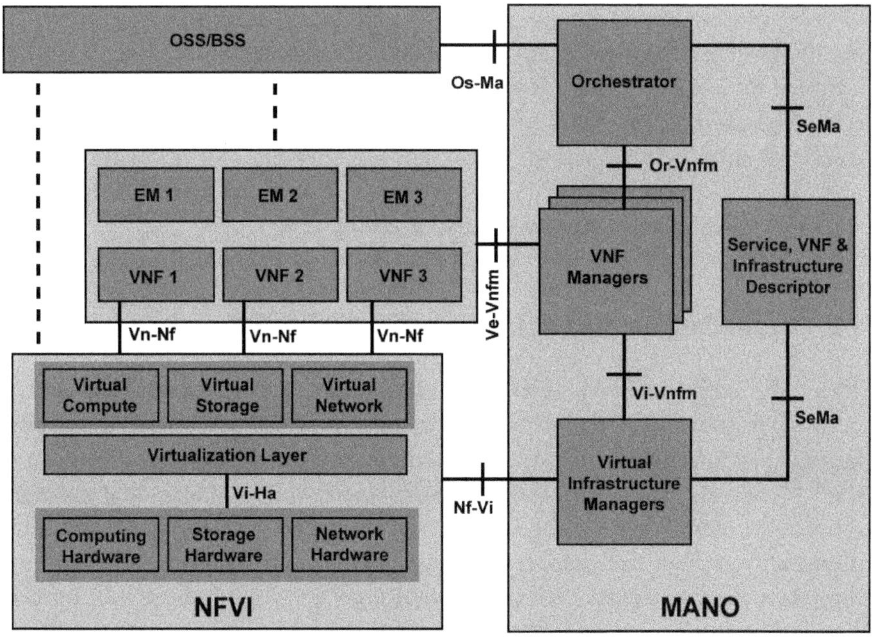


**Figure 1.1** Introducing NFV for the first time.

the legacy EMS was often a vendor-specific separate system or set of systems that was used to monitor and manage the network functional elements. This would include collecting performance and log or alarm data from the integrated system. Today these EMS functions rarely are part of the NFV platform and instead reside on other nodes in the network, either with MANO functions or on other elements. The MANO too has evolved from the original and continues to evolve today to meet the needs of the operators and developer community.

Since the time of writing, there has been significant development work and additional abstraction to this design. Today we find the basic model is well applied and significant progress has been made to the underlying concept of separating the functionality of hardware from software for solutions in the telecommunications industry, which is currently called disaggregation.

Today the idea of disaggregating the network functions from the underlying hardware to achieve the first stated goal found in the white paper, which is "reduce equipment cost and reduce power consumption," is having a positive impact on the capital expenditure (CapEx) (the cost of adding assets or adding value to existing assets; these expenses are depreciated over time) and operational expenditure (OpEx) (the costs incurred to run and operate a business) for the network operators. The NFV paper also emphasized a goal of reducing the time to market (sometimes called agility), both for bringing existing services into scope and introducing possibly new features and services. The goal of continuously improving the speed and agility of the network operations remains as much a focus today as it was when NFV was first envisioned. It is common for the disaggregation of the network functions to be spoken of in terms of virtualization. Indeed, it is fair to consider these terms synonymous in many cases; the intended goal was the separation and then later recombination of the underlying hardware and the upper layer application along with the supporting middle layers and associated system integration.

## 1.2 Expanding on the First Vision of Virtualization

There will be more on the evolution of virtualization in the communication network in later chapters, but for now we return to the opening statement. To begin the expansion of the opening statement, let's dissect this from the right to the left, starting with what a communication network is. Generally, the communication network should be considered the global network that allows the ubiquitous exchange of data in multiple forms in near real time. If we accept this, then the commonly well understood internet is a subset of the communication network. One view shown in the center of Figure 1.2 provides an abstraction of the communication network. In the center, the telecommunication network reaches close to the edge where the users (consumers and enterprises) connect. This reach may take several physical forms, which include both over the air and those with physical media (e.g., fiber or metal wires). There are other

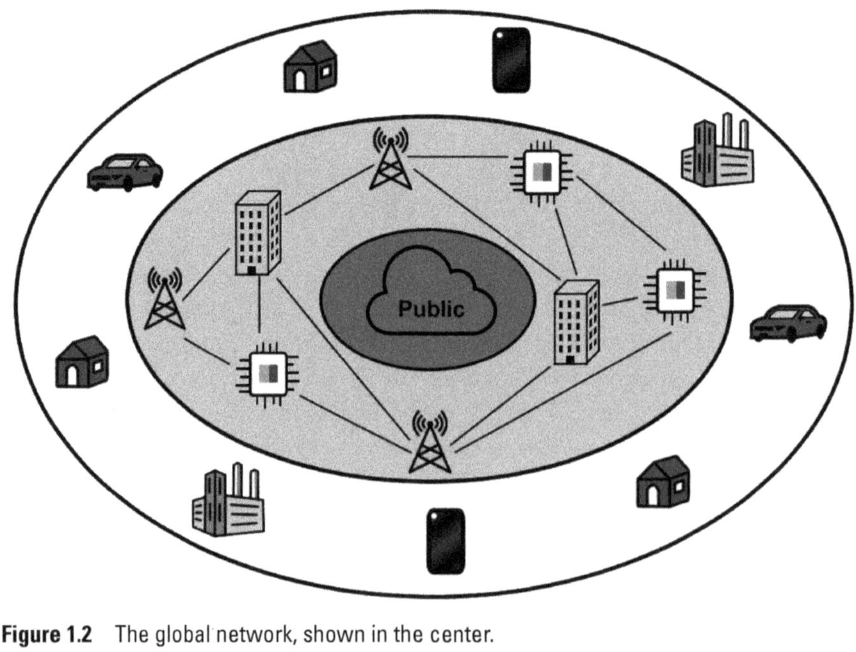


**Figure 1.2** The global network, shown in the center.

networks that we will not be discussing, which may be multinational in scope or controlled by national entities. For simplicity, the network we are discussing here is the global network that is still capable today of connecting two people anywhere on the globe who have a phone number and/or broadband connectivity. This of course is not the full extent of the capabilities of the telecommunications network, but rather a foundational element that is still supported today.

This network consists of both wired and wireless segments, and while both are undergoing the transformation to virtualization it is the mobile network (also known as the wireless network) that has proven to be the leading candidate in this space. The global communication network is made up of well over 600 mobile network operators. These operators are defined here as those that hold a license to use mobile network spectrum and offer services that rely on the utilization of that spectrum as a part of their business. There are an even larger number of network operators that may not have license for spectrum, and others that use the spectrum of the licensee. This global communication network is comprised of the equipment that allows both people and enterprises (e.g., any entity or thing that is not an individual will be referred to as an enterprise, which includes different types of governmental and nongovernmental users) to exchange data over a common infrastructure. The variety of methods of data exchange, the rates, and the format is large, and often the network doesn't care about the format. The network that realizes the mobile portion of this global network originally evolved from the legacy phone network where the data exchanged was founded in the requirement to exchange human voice in full duplex (e.g., both parties could be talking simultaneously). The network continues to support this fundamental capability today, the ability to carry a voice call in full duplex. The network today carries a significantly greater volume of data that is not voice calls; nevertheless, they are still an important aspect of the design considerations. Additionally, the ability to carry voice calls and the service uptime may be heavily regulated by the license granting authority (e.g., the ability to call emergency services over the network).

## 1.3 Breaking Down the Fundamentals Driving Virtualization

Returning to the opening statement, the engineering and operation of the communication network touches both the CapEx and OpEx of the network. CapEx in this case is the accumulation of the cost paid by the operator to purchase and install all the materials associated with building the network infrastructure. This includes the purchase cost of any systems (hardware) and in many cases the initial licensing of any associated software and may also include the cost associated with installing ("racking and stacking") of the systems into a specific site. The OpEx cost is the recurring costs of running the network, this would include the cost of power (power to operate the systems and the cost of cooling) and any direct labor cost associated with maintaining the operation of the systems in the network, along with any recurring hardware or software maintenance fees. The vendor of the hardware and/or software often includes some level of ongoing support, also known as maintenance, and on an annual basis requires the operator to remit payment for the next year's maintenance. There is a new category that is starting to emerge in some markets and by some vendors that is solely based on collecting an ongoing maintenance fee, deferring the initial and usually larger CapEx, and relying solely on an OpEx model of continuously recurring usage fees. In the case of CapEx, the operators can often take investment credits on their tax obligations in some jurisdictions where the OpEx often does not qualify for this accounting. While both expense types represent a significant expense to the operators, we often see models where the trade-off of one expense verses the other can change the rate at which an operator is able to introduce new technology into their network. This case arose with the introduction of 5G and will be discussed later when the topic of standalone and non-standalone 5G core is discussed.

By some estimates, the total cost to operate a modern system in today's network is nearly evenly divided into thirds. Unfortunately, data of this nature is closely protected by the operators, so our estimates are based on the distillation of several conversations protected under nondisclosure agreements between global communication service providers (CSPs) and one of the authors.

The first third is the CapEx to bring the function into the network. The second third is the power to operate and cool the systems during their operational life. The final third is the operator's labor and recurring maintenance fees associated with the solution. One third of the total cost of ownership (TCO) is CapEx and the other two thirds of the TCO are OpEx. There are a few geographies where abundant hydroelectric power may have a significant impact in reducing the energy cost in the OpEx portion, but these are rare, and one might think of areas vast in land and sparse in human population to find examples where this holds true. In addition, it is estimated (again, usually a closely guarded metric) that some network operators with both wireless and wireline networks allocate 70% of their CapEx to the mobile network, and as a result a fair share of the following OpEx. With this foundational understanding it should be clear why these network operators would be very interested in pursuing the goal of the 2012 white paper as pointed out in the opening.

## 1.4 Applying This Discussion to the Mobile Radio Network

The foundation of applying this discussion to the virtualization of the 5G and Beyond 5G (B5G) mobile network brings the opportunity to introduce to those new to this subject the latest on the work as well as to both the technology and the industry. For those well versed in either the current technology and/or the business motivation, we hope to stimulate additional work on the transformation ongoing in this global network.

## 1.5 Transforming the Mobile Network One G at a Time

One very common trend when discussing the mobile network is to describe the steps in terms of the Gs (generations), 1G, 2G, 3G, 4G, 5G, and Beyond 5G. Here 1G is the first generation, which was not called 1G initially—readers with long memories may still refer to this as Advanced Mobile Phone System (AMPS). Here we simply give it the first counting number for convenience and find little reason to define or discuss 1G except to simply reference it as the starting point for the network we have today.

This progress of Gs is shown in Figure 1.3.

The caution here is that while the industry speaks of the Gs incrementally, there have also been many sub-Gs along the way, as shown in Figure 1.4, and network operators often have multiple variants active in their network simultaneously, so that in many cases the instance of more than one G can be realized on the same hardware with other Gs at the same time. This has sometimes been described as "stretching the metal" around multiple G specifications at the same

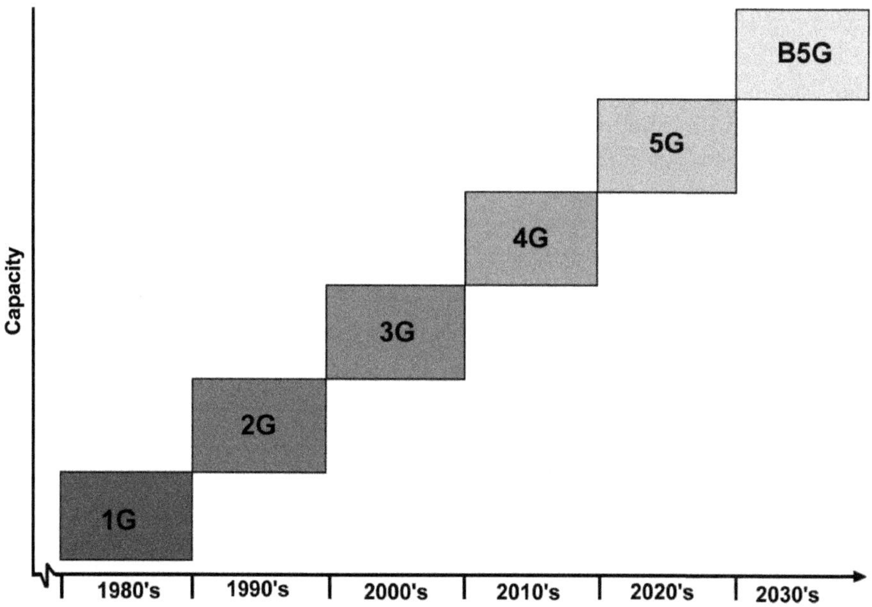


**Figure 1.3** The Gs of the mobile network.

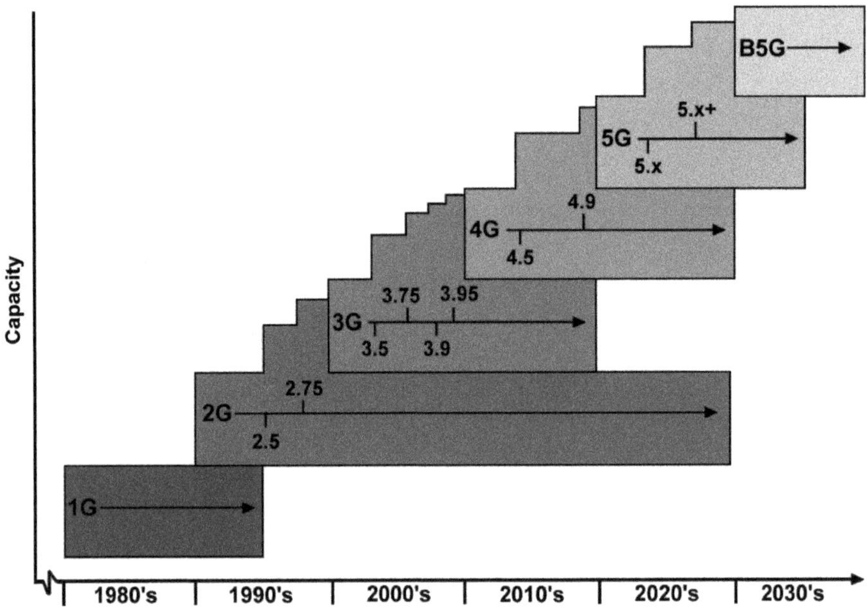


**Figure 1.4** Evolution of the mobile network versions.

time (the computer chassis are made of metal, and if one collection of these computers are running, for example, 4G and 5G on the same systems, we've stretched the metal across 4 and 5G). Said differently, for example, a Mobile Management Entity (MME) function (MME is an element found in the 4G network specification) in 4G may be realized on the same computer hardware as the access mobility management function (AMF) for the 5G network (for now MME and AMF can be thought of as an application workload on a computer; there will be more to say about these functions in later chapters). One may presume simply from the expansion of the acronym that these two might have some commonality in function and in fact they do, and this is often the case from one G to the next; there is an evolution of functionality and often a name change in what are otherwise similar modules. With 5G we find better separation of functions that allows for the virtualization (NFV and SDN) design considerations to be realized in the software.

5G has three well defined enhancements above the capabilities found in 4G, which are shown in Figure 1.5: enhanced mobile broadband (eMBB or EMBB), ultra-reliable low-latency communications (URLLC), and massive machine-type communications (mMTC). The terms eMBB, URLLC, and mMTC are preferred, but from time to time we find that some authors may choose to use all caps to align with the URLLC format. 5G was specified to support use cases where massive bandwidth and very-low-power devices could

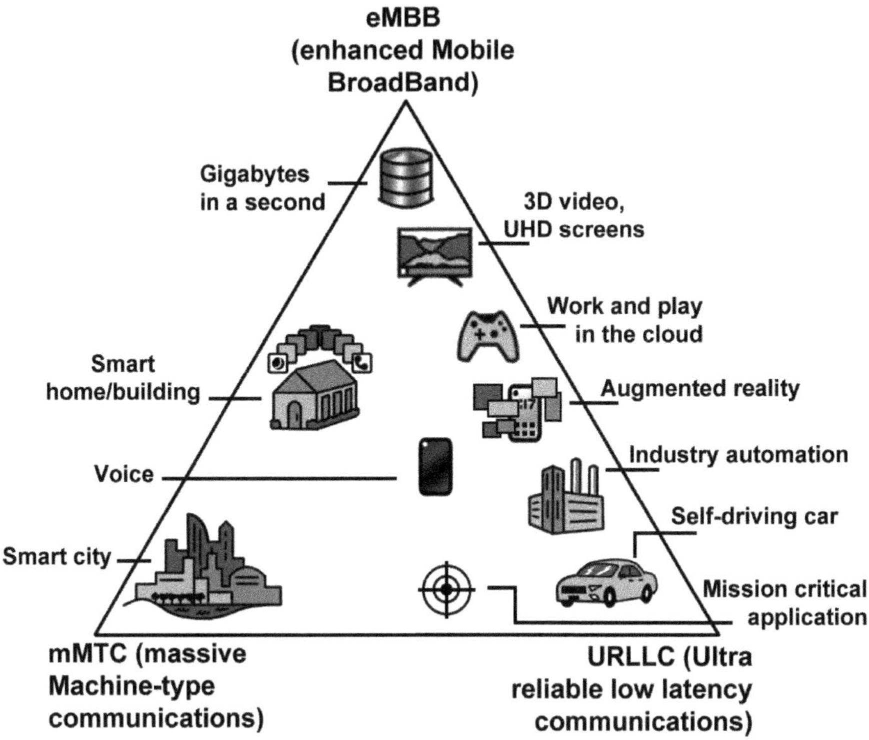


**Figure 1.5** 5G feature drivers.

coexist in one technology, but not necessarily on the same user equipment (UE) device. Design goals speak of up to 5 years or more of battery life for some devices, and, at least on paper, up to 10 Gbps download speed for other devices; the 10 Gbps of 5G is 100× the "on paper" 4G download speed. Additional design considerations in the 5G specifications allow for the separation of network user traffic and management functions in addition to the ability to migrate from 4G to 5G while reusing portions of the 4G network. 5G also introduced concepts allowing traffic steering where not all traffic would receive the same treatment. Some traffic may be processed closer to the source, often called the edge, while other traffic from the same origination point might be transported deeply into the network. Previous generations of the Gs required that all data be treated exactly the same way. This opens 5G to wider use cases beyond the consumer space, and when coupled with innovative spectrum licenses in some jurisdictions, has the potential to open new markets and drive innovation.

The reality is that the evolution of the G has taken several smaller steps from time to time, a more realistic view of this stepwise evolution. Many historical network operators have been running mobile networks since at least the 2G era, and the reality is that once hardware systems have been deployed in the network, for example at the start of the 2G rollout, these systems might not be capable of running the 3G workloads. The operators and the supplier ecosystem then had two approaches when 3G rolled around and the 2G network was (and in many places still is) operational. Like the sides of a coin choosing heads over tails or tails over heads, there were simply two options to consider without clear advantages one way or the other: continuing to operate the 2G network equipment and bring in new equipment to operate the 3G network, or when bringing in the 3G network equipment ensuring that it was also capable of running the 2G workloads.

For example, choosing to deploy separate systems for 3G would increase the number of computers that must be managed in the network and may require more rack space in the installed location, and also possibly require more power and interface ports. But while upgrading the system to a single platform that could support both 2G and 3G may bring some advantages in space and power savings while reducing the number of nodes that need to be managed, it would possibly incur increased hardware cost. In this case, the operator may have already purchased the 2G hardware and licenses and might have to repurchase some of the compute resources as a minimum.

## 1.6 Evolving Small Steps on the Gs

There are several points of interest when comparing the differences in Figures 1.3 and 1.4. First, the introduction of the next G does not terminate the previous G in the network, noting that 2G appears to have a longer life than 3G. At the time of writing, some network operators are terminating their 3G and will likely repurpose the spectrum for usage in 5G deployments. Briefly, a mention of where spectrum comes from may be required; this is a portion of the electromagnetic spectrum known as the radio frequency (RF) portion. Each nation claims sovereignty to this limited resource and in many cases issues licenses for the use of specific bands or portions of the spectrum. Often these licenses are obtained for a fee. Just how much of a fee depends; for example, according to a recent article in *IEEE Spectrum,* the total value of the spectrum auction in the United States over the past 30 years was about 230 billion USD for all licenses. This includes not only the mobile network spectrum but also other licenses as well, including TV, satellite, classical radio, and others. Clearly, spectrum has high value for both the licensing jurisdiction and to the licensee [3].

A keen eye looking at Figure 1.4 will notice that 2G lingers past the 3G timeline. Why keep 2G around longer than 3G? 2G has a number of advantages to some legacy enterprise use cases and while the typical consumer may no longer be a 2G user, it is likely that it will be many more years before all usage of 2G expires. Regulators may also play a role in the decommissioning of a particular service; operators were not able to just turn down their 3G and abandon subscribers who had not upgraded their devices. In many cases regulators required operators to provide compatible devices to these 3G consumers at no cost to the consumer. In some jurisdictions, portions of the CSP network, if not all if it, are considered critical infrastructure and as such, regulators have significant say about how a business can be operated and they take regulatory compliance very seriously.

There are a number of sub-Gs as shown in Figure 1.4, such as 4.5G and 4.9G. In some cases, equipment installed at the beginning of a G's timeline may not be capable of running the last iteration of the latest G. Consider the case when two Gs are intended to be run together on one set of systems prior to the actual turnup of the latest G. A set of systems in the network currently running 3G, for example, may have to be upgraded prior to the introduction of 4G into the network if these systems are intended to support both 3G and 4G. As time passes and new subvariants of 4G emerge, it can be possible that this equipment is incapable for a number of reasons from running the latest iteration of 4G. In this case the operator has to either install new equipment for the latest sub of 4G, or forgo that upgrade and risk that competitors will outcompete them. For example, systems installed in the late 2008/9 time frame to support both 3.95G and the soon-to-be-turned-on 4G may not have been capable of running the 4.9G release in the late 2010s. This is the crux of the problem network operators were facing—these purpose-built network computers were constantly requiring upgrades to support the current technology in the network and positioned for the next release. This, and even more significantly the ever-increasing traffic load on the telco network, required that the operators take a new look at how they build, deploy, and operate their networks, hence NFV. Telco, for those unfamiliar, is a nickname for a telecommunication service provider, and is often used to refer to a class of service providers today, providing at a minimum connectivity into the wider global communication network. Telecom is often used in a wider sense than telco, and is used to refer to the wider ecosystem rather than just to the service providers themselves (e.g., a telco is an operator, telecom is an industry and includes more than just the operators themselves).

## 1.7 Which Network Is This Exactly?

Some readers will already know that the terms "telecom" and "communication networks" are sometimes used interchangeably. This is intentional because "telecom" can be a noun for an industry and "communication networks" can be a qualified noun for the physical realization that enables a business to operate in this domain. Webster's [4] provides some insight here. While it is expected that the concept of a network is understood by anyone interested or involved in this area, new entrants into this field may benefit from a more thorough treatment.

## 1.8 Acronyms and Domain-Specific Terms Abound

Telcos, CSPs (also known as CoSPs by some silicon venders in their literature), network operators, operators, carriers, service providers, multiple systems operators (MSOs, also known as cable operators), and a few other terms are used as names for the same thing. While we will strive to be consistent in our usage in this book, note that different regions of the globe use different terms for this industry and for the companies that are generating revenue from the use of the communication network. Furthermore, even within some communities, the names are used interchangeably. As an additional note of caution, some entities have chosen to abbreviate this even further: the acronym CSP is common in some press and publications covering this industry, while at least one large influential silicon manufacturer prefers CoSP, reserving CSP for cloud service providers (also known as hyperscalers like Alibaba, Amazon, Azure, Google, etc.). We will continue to use CSP to mean communication service provider, as it is more generally accepted at this time. At times, however, other terms for the telcos or CSPs may be used, such as "operators," where either the context and flow fit the usage better, or where they can serve as a reminder of the variations that will be encountered in the industry at large. There is an entire ecosystem of solution providers into this industry that are also known by various names and associated acronyms that are used interchangeably, such as NEPs and TEMs (NEP is generally preferred but TEM still can appear in the literature). To determine the context of the three-letter acronym (TLA) or term in use, the reader may from time to time have to consider the scope of the material. One can consider learning the various overloaded synonyms that will be encountered as part of the criteria for entry into this field.

## 1.9 Telecom Providers Go by Many Names

Terms like mobile network operator (MNO) and mobile virtual network operator (MVNO) will also be encountered in the literature, presentations, and conversations. Today, most MNOs also have wireline business segments, and our hope is that this term begins to fade as we view the network as a unified entity. One last note of caution in the naming conventions—the V in MVNO is not "virtualization," which refers to the business of reselling possibly rebranded service on an existing operator's network. In some markets, MVNOs evolved to provide a lower-cost service possibly based on a prepaid subscription model, while in others it was market brand driven; for example, standing up a wholly-owned subsidiary with a younger generational name and possibly a different level of service than found on a premium-branded offering.

These discussions will primarily center around the businesses and technology that have evolved from the legacy wired communication network. There are two other domains of existing networks that are now intersecting with what can be called the historical mobile network operators: the cable and satellite operators. The intent is to include them in the discussion on green-field entrants into the mobile operator space rather than attempt to include the evolution of their networks to grow into the traditional telco space. This is not to discount their evolution or their contribution to the global communication network, but to confine the overall volume of this work to a manageable size. These new entrants into the mobile communication network as defined here have evolved their networks with different technologies and with a different cadence than that of the traditional mobile network operator. While the underlying services have similar capabilities today (namely the transportation in near real time of end-to-end data including voice), the names of the nodes, the protocols used, and the vendors supplying equipment into their business sector are different. Their entry into the MNO space will very likely see their network elements (at least for the mobile portion) align with the standards bodies and ecosystems we will be discussing. The final new entrants as discussed above are those coming in without a legacy network of any kind. We have seen three to date, one in India entering initially as a 4G provider, one in Japan entering as a late 4G and early 5G provider, and one in the United States coming from broadcast satellite. Each of these entrants came into the space with an initial vision to virtualize as much of their first network as possible at the time, bringing the cost and efficiency savings envisioned in the 2012 white paper from the start.

## 1.10 Addressing the Various Audiences

This book is intended to aid new entrants into this domain (e.g., students), practicing engineers (those developing, designing, building, and operating networks), members of business teams working in this domain, and academic researchers. It will provide fundamental and theoretical insights as well as hands-on discussion for these practitioners. For those new to the communication network domain, some introduction to this industry is presented in early chapters. A seasoned veteran of this industry will find some sections useful and possibly gain fresh insight. The review of where this industry has evolved from may aid in seeing the most likely vector toward the future. As we have shown, this industry is rife with TLAs. In addition, there are other abbreviated terms and borrowed concepts that are deeply rooted in the day-to-day language used and at first this might be a distraction to the newly initiated. To understand the full context of the material being discussed, every attempt will be made to provide context and clarity when a new term or concept is introduced. Not all concepts or terms will be fully defined here due to the limited space available. See the discussion in the Glossary of Acronyms and Common Terms for more on this subject.

## 1.11 To Those New to This Industry

Welcome to telecom! This industry is very much like an ancient forest—there is much to explore, some of it very old and in good condition, some possibly in a more declining state, and even new life if you are willing to have a deeper look. Virtualization represents some of the new life in this old forest.

Our goal is to introduce the reader to the technical and business fundamentals driving the virtualization of 5G and B5G in mobile networks.

## 1.12 Structure of the Remaining Chapters

This book is organized into three major sections with multiple chapters in each section. Chapters 1–5 cover the fundamentals, Chapters 6–11 cover engineering considerations, and Chapters 12–14 cover future developments. The book ends with a list of acronyms and nomeclature used in the field.

### 1.12.1 The Fundamentals: Chapters 1–5

The remainder of this chapter continues with the structure of the book and provides guidance on the usage of the subsequent chapters.

*Chapter 2* discusses the benefits of NFV for 5G and B5G networks. We begin with a holistic and techno-economic discussion of the anticipated benefits when using virtualization techniques in mobile networks. We address what the technical and business case is that motivates the transformation that comprises NFV, and why is this being discussed in the 5G use case and B5G? The discussion of virtualization technologies begins with a discussion on the acceleration of packets in the data plane using SDN techniques.

The historical fundamentals of virtualization are presented briefly, leading to a discussion on how the performance of virtual mobile networks might be achieved and eventually differ from nonvirtualized networks, and how to measure a successful design of virtual 5B/B5G networks. The chapter concludes with a discussion on the business impact and increasing interest of private 5G/B5G networks for industrial users and outlines their implementation and application pathways.

This chapter will provide the reader with a solid foundation on understanding the why in addition to the what and how, which will be covered in the subsequent chapters.

*Chapter 3* discusses the current state and practices in virtualization of the CSP network. This includes coverage on the interesting workloads in the network. The control plane and data plane are discussed along with the their challenges; for example, the control plane is stateful, and the data plane has significant performance requirements (e.g., throughput and latency). To address the needs for performance, acceleration of the data packets is required, which takes the conversation to field-programmable gate arrays (FPGAs), smart network interface cards (NICs), and new languages such as P4 for smart switch hardware.

Also covered are data collection and analysis to measure and monitor the server performance metrics.

This includes virtualization of the CSP workloads, core network design, control plane virtualization, data plane virtualization, and packet acceleration. Within packet acceleration both hardware and software acceleration techniques are covered, and includes use of FPGAs, eASICs, and programmable ASICs. eASIC is a new class of ASIC with some limited ability to be configured after initial manufacturing (e.g., at power up). One key point is the ever-present concern over the power budget (both in terms of kilowatts-hours consumed and the actual cost). The role that power consumption plays in the network is extremely important, both technically (e.g., dissipating heat) and economically (e.g., energy cost).

*Chapter 4* explains that data plane virtualization consumes the most significant portions of the operational system and there are techniques and trade-offs that must be considered to maximize the performance and minimize the cost, both in capital hardware and in the power consumed in operating the systems. This chapter covers the current implementation options available to meet the demands of the 5G network. One technique is to use SDN to bring control and flexibility to the network operations.

Hardware acceleration remains a critical component in the network today. This can be realized in several ways: NICs, smart NICs, separate FPGA cards, programmable ASICs (also known as eASICs), and modern switching fabrics that are programmable using the P4 language. The chapter concludes with a detailed discussion on P4. Specialized NIC interface software techniques and packet acceleration in the core with techniques like DPDK (today it's just DPDK, but expanded it was Data Plane Development Kit [5], single root I/O virtualization (SRIOV) [6], and eXpress Data Path (XDP) [7]) are covered in later chapters.

*Chapter 5:* Performance remains a key topic in virtualization. Drivers include how TCO is minimized while providing a consistent SLA. In this chapter the key performance measurements are mathematically defined and how they are determined in a virtualized network element are considered. CPU pinning (ensuring a fixed workload is anchored to a fixed CPU core) is discussed. Non-uniform memory access (NUMA) awareness both at the server and within a CPU are also covered. In addition, how this might differ from a cloud workload is considered. Developers and users of the network require deep insight into the performance of the systems. This requires instrumentation that enables detailed measurements of the various elements of the system, from the hardware up to the application layer and may include system controls that are aware of workload placement. This is important because CPU pinning (e.g., the ability to specify exactly which core in a die a threat is required to be executed on is fixed and it cannot be moved or rescheduled to another core on the CPU) can have implications to network traffic engineering that requires NUMA-aware [8].

### 1.12.2 Engineering of Virtualized 5G and B5G Systems: Chapters 6–11

*Chapter 6:* Many mobile operators today have significant investments in 4G network infrastructure, and the introduction of 5G in many cases will overlay with existing 4G networks for the foreseeable future. In some cases, even the nodes that are deployed will support both 4G and 5G functionality in the same system. Transforming these systems to a pure virtualized network will take considerable effort from both the NEPs and the telcos.

Nearly every operator today has legacy networks and thus teams of engineers practicing in this space that use or have built existing internal systems. Thus, there is internal competition that bring what some have described as "institutional antibodies" who resist any change.

*Chapter 7* goes into the design and implementation of the modern virtualized RAN, in particular the 5G RAN. The standards from 3GPP [9], the main standards body, and the ORAN alliance [10] and vRAN design principles will be covered in detail. Splits (partitioning) in the design will also be covered in this chapter. There are a number of possible considerations when building the virtualized RAN, one of which is the choice of splits that may be deployed. There is no one design, but rather a large number of possible configurations driven by deployment considerations. Some of these are based on geographical considerations, while others may be driven by power and space considerations. This chapter will investigate the various implementation requirements in depth.

*Chapter 8* will show how the metal is stretched around the specifications found in the standards. We have already introduced the concept of stretching the metal, and we expand on that in this chapter. Stretching the metal refers to two things. First, the computers that realize the various functions in the network are cased in metal chassis and the silicon devices (fixed function or programmable) are constructed using metal. Thus, both the computer hardware and chassis have metal stretched around them. Second, it is an expression of the combination of functions from various generations (e.g., 4G and 5G functionality) realized in hardware or software that in some cases may be found executing on the same system. Thus, we have stretched the functionality of a system to include multiple functions and possibly even within the same generation.

While much of the discussion in this work spans the 5G network, not all networks are pure 5G. In some cases, an analysis of components reveals that real networks today may support 4G, 5, and Wireline all wrapped into one. Many network operators and network equipment providers have 4G functionality running within the same software and on the same hardware as 5G elements. This is important for several technical and business reasons. In this chapter we will look at the desire to have standalone and non-standalone functionality in the network, or, what can be referred to as "how the metal is stretched around the specifications."

*Chapter 9* expands on the work in the previous chapter and breaks the RAN into bite-sized pieces. The design and implementation considerations of the various splits will be discussed in terms of both technical and business concerns. The focus on the disaggregation of the modern RAN and how design splits are realized are discussed. The chapters also examines business considerations in virtualization as it implies changing the business model of how the networks are built and operated.

*Chapter 10* addresses the question "How does the industry measure success of virtualization?" There are multiple cost drivers. One is the cost of developing and maintaining software. Another is the OpEx associated with operating the network (e.g., how many people are required to maintain the mobile network). To address the first cost driver, the chapter gives particular attention to open-source software (OSSW), its capability to reuse software development and maintenance efforts, and available 5G. The chapter also includes a discussion of performance and operational criteria for 5G systems.

*Chapter 11* looks at how operating the network at scale is a significant concern to mobile network operators. To tackle and manage the growing complexity with scaling, it is critical to automate the process of deploying, optimizing, and operating the RAN while also taking full advantage of newly available data-driven technologies (for example artificial intelligence (AI) and machine learning (ML)) to ultimately improve the end-user quality of experience. The RAN intelligent controller (RIC) is an area of significant modernization in today's RAN. This chapter will discuss disaggregation, RIC, security, support, and the lifecycle management of the nodes and operational considerations of the virtualized network. One key consideration in that operation and scale includes reconfigurability and migration of virtualized modules.

### 1.12.3 Future Developments: Chapters 12–14

This part provides a more forward-looking view of what might be coming in the industry as we continue to roll out 5G and begin considering the implications of what comes next.

*Chapter 12* details how 5G brings a new market potential to the telcos. One possible area is in private networks for enterprises—a potentially significant opportunity for revenue generation for the telcos. This chapter will cover the technical and business opportunity at the "the edge" where one finds the private networks. One significant challenge here for the CSPs is to expand past their legacy view of providing connectivity and security to the enterprise, and pivot to providing domain-specific computing at the edge. The current trend is that the services of private networking are being led by the NEPs, the systems integrators (SIs), and the enterprises themselves, with the CSPs currently providing only 20% of the overall private network deployments.

*Chapter 13:* The scientific community at large has responded to the many challenges of designing and developing high-performance 5G technologies in two ways. On the one hand, it has focused on the development of open-source 3GPP software implementations and deployment frameworks that can be effectively used to quickly build 5G proof-of-concept (PoC) designs. On the other hand, it has pursued the construction of publicly available open experimental labs that offer researchers time-shared access to various 3GPP technology services. These ongoing efforts are reviewed in this chapter.

*Chapter 14* offers final thoughts on the current and possibly future state of the effort to disaggregate the operational mobile network. In addition, new challenges that remain unaddressed, continued packet acceleration capabilities, better power management, improved cloud native operational considerations are put forth.

### 1.12.4 Acronyms and Terms

This section lists many of the acronyms that are part of the common language—the alphabet soup—found in the communication industry. There is a very long list of TLAs and terms that have been abbreviated or used extensively in the industry, and by no means is this an attempt to list all that may be encountered. A more complete list can be found in *Telecom Dictionary,* which is still available from some sources [11, 12].

## References

[1] https://portal.etsi.org/NFV/NFV_White_Paper.pdf.

[2] IEEE Conference on Network Function Virtualization and Software Defined Networks, https://nfvsdn2022.ieee-nfvsdn.org.

[3] "Billionaires Battle for Global Spectrum," *IEEE Spectrum*, October 2022.

[4] https://www.merriam-webster.com/dictionary/telecom.

[5] https://software.intel.com/content/www/us/en/develop/topics/networking/dpdk.html.

[6] https://docs.microsoft.com/en-us/windows-hardware/drivers/network/overview-of-single-root-i-o-virtualization--sr-iov.

[7] https://www.kernel.org/doc/html/latest/networking/af_xdp.html.

[8] https://en.wikipedia.org/wiki/Non-uniform_memory_access.

[9] https://www.3gpp.org/about-3gpp/about-3gpp.

[10] https://www.o-ran.org/.

[11] https://www.barnesandnoble.com/w/newtons-telecom-dictionary-harry-newton/1119676342.

[12] http://www.telecomdictionary.com/index.asp.

# 2

# Benefits of NFV for 5G and B5G Networks and Standards Bodies

## 2.1 Why Use NFV for Networks?

This chapter explores the business and technical question: Why is the industry changing the way the networks are built and operated today, and in doing so, contrasting this new approach with the previous approaches? The technology that exists today transforms the legacy structures to a large extent, and this can improve the business model the telcos use to build and operate their networks. An understanding of the "why" in addition to the "what" and "how" establishes a firm foundation for the following chapters. Also covered here is the role of several significant standards bodies in contributing technical requirements that allow for the global interoperation of the mobile network we know today.

### 2.1.1 Transformation of a Large Legacy Business Is Difficult

There are significant areas that slow the transformation of the legacy communication service provider community. These areas lie in the vendor community and within the telcos themselves. One is inertia both in the existing ecosystem and in the telcos and a second is the difficulty of innovation in the industry. New entrants into the industry face significant technical and financial barriers to entry: long development times, longer procurement cycles, and strong vendor relationships that go back decades in some cases with the telco customers. Nevertheless, there are existing examples where new entrants have penetrated the market, bringing innovation and new business models to the industry. There have also been new entrants into the service provider space themselves, and while rare, they have proven that the industry can innovate both in how the network is built and operated and how they brought disruptive vendors into the industry.

## 2.2 The Existing NEP Ecosystem of Vendors

The existing ecosystem of the TEMs (also known as NEPs) comprises a very limited group of very large companies (see Figure 2.1).

While Figure 2.1 gives the total revenue for each for a few years back, the significance is that with many above 20 billion USD in annual sales these by no means are small companies. It should also be pointed out that several are well over 100 years old (in order of their founding: Corning 1851, Nokia 1865, Ericsson 1867, NEC 1899, Fujitsu 1935, Cisco 1984, Qualcomm 1985, ZTE 1985, and Huawei 1987).

The revenue profile of each of these companies is not within the same domain; for example, Corning has a large fiber and glass product line and

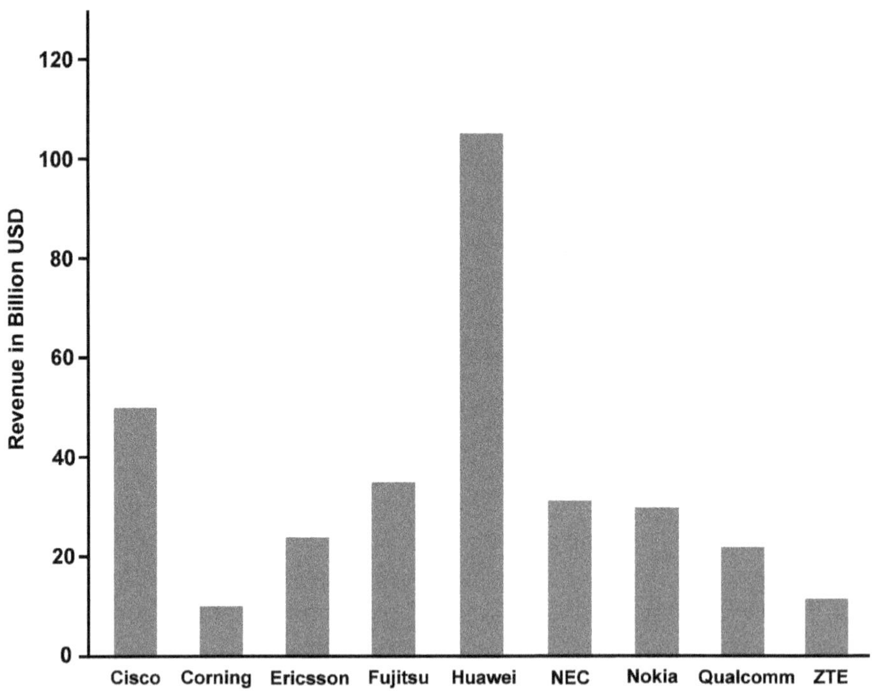


**Figure 2.1** Revenue per top CSP supplier (NEP/TEM).

pharmaceuticals along with other business areas that are not replicated by any of the other eight. Qualcomm likewise has a user equipment (UE, e.g., a mobile handset) modem business unit supplying the largest share of handset modem chips. One might also ask why companies such as Samsung and Apple are not included. This table was generated from a query of the top revenue-generating companies in telecom equipment manufacturing. The other two, while very large indeed, do not generate as much revenue in supplying equipment and services into the network operators directly. We will see some of the Apple and Samsung impacts represented shortly.

These ecosystem partners in some cases have been building and deploying equipment and solutions into this industry for over a century. Their business has grown to include the ability to not only build the equipment that is consumed by the network operator, but to also perform systems engineering and operation of the system once installed. In addition, some of the TEMs provide complete lifecycle management of the systems once installed. This work would include providing and applying software upgrades and if a piece of hardware fails the TEM is engaged to replace it. Software upgrades may include enhancements that cure defects, close security holes, or introduce new features developed by the TEM. These business relationships are normally covered under maintenance contracts where the CSP (or CommSP, also known as the telco) is paying an annual fee to the TEM for this service in addition to the normal ongoing software license and hardware maintenance costs.

## 2.3 Changing Business Models Midstream

One of the transformation challenges lies in the size and organization of these large and complex companies. This of course is the challenge many technologies companies face—how to transform an existing business model and remain viable during the process. A prime example is the former U.S. company Eastman Kodak (Kodak) founded in 1892. We stepped outside the communication field so as not to cause too much disagreement here. For those unfamiliar, Kodak for most of the 1900s was the dominant photographic film company. Kodak developed the first self-contained digital camera, and yet in 2012 filed for bankruptcy. Today it is a shadow of its former self. Kodak could not transform from selling chemicals and film quickly enough when the digital transformation hit them. This is not an attempt to predict any future in the CSP industry, but rather to point out that technology does not wait for market trends to be favorable. Disruption forces can exist inside a company and there still may not be enough forces to counteract internal resistance. One colleague calls this being attacked by the institutional antibodies.

One of the companies on the list above, Cisco, is a counterstudy in their ability to transform and direct resources and thrive as a result. Cisco in the late 1990s and prior to the internet bubble burst in March 2000 used a model of acquiring a number of promising startups and quickly integrating their technology and staff into the Cisco products and culture. This method allowed Cisco to direct their research and development (R&D) funds to their existing products and markets while nascent technologies were developed and tested out by startups, with Cisco then acquiring the startups at just the right time for the Voice over IP (VoIP) era.

## 2.4 Independent Software Vendors as NEPs

Today there are a number of growing independent software vendors (ISVs) that have had some market success in the CSP space, and yet there is over little overall investment in new startups targeting CSP networks and operations and the mobile network in particular. One key reason is the very long lead times to develop product and existing vendor lock-in with CSPs. There are, however, three large-scale new entrants in the mobile 4G/5G space. These completely new entrants as service providers in major markets in the past decade are Reliance Jio in India, Rakuten in Japan, and Dish in the United States. We introduced the term "green field" in Chapter 1; for those unfamiliar, green field refers to a new entrant into a business or industry that has no legacy customers or legacy infrastructure to support [1]. Green field can be easier to transform (both hardware systems and staff), as one is starting without any existing technology, staff, or procedures that need to be supported and transformed.

## 2.5 Green-Field Entrants into the CSP Business

Jio stood up a 4G network without any previous generations to support and brought an innovative market-winning approach to attract customers in their first few years of operation. This was accomplished with a significant investment that in two years grew from zero to over 130 million subscribers.

Rakuten entered the market in Japan with a green-field 5G network built with the largest proportion of virtualized network possible at the time, and has subsequently acquired one of their RAN providers, Altiostar. Rakuten remains an interesting transformative force as they have the ability to license their technology and 5G solution to other CSPs.

There is a third entity that is quasi green field. Dish Networks is a satellite service provider that is turning up a terrestrial 5G network, and while Dish has some capabilities that might be associated with a traditional CSP, we would view them as also coming in as green field because they have no land-based legacy or prior Gs to support.

There are several more modestly sized ISVs developing and delivering solutions into the CSP 4G and 5G RAN today, two of which were acquired by Microsoft—Affirmed Networks and Metaswitch in 2019. The hyperscalers clearly have interest in the CSP markets and are developing and acquiring technology and staff knowledge in this space. Their largest barrier today is access to spectrum licenses. Spectrum licenses will be discussed in more detail later.

## 2.6 Transformation from Hardware-Centric to Software-Centric Networks

Returning for the moment to companies in Figure 2.1, a decade or so ago each of these, possibly with the exception of Corning, were vendors of purpose-built hardware solutions. A CSP would procure a particular solution from the vendor, and the systems that realized that solution would have been branded with the logo or name of the vendor. Today, while some still offer physical devices (servers possibly with enhanced capabilities), many also offer at least a portion of their solution as software only that has been validated to perform well and is supported on specific servers from original equipment or design manufacturers (OxMs; the x means either equipment or design). These would include brand names like Dell, HPE, SuperMicro, Silicom, and Lenovo, and are more commonly known as industry standard high-volume servers or white boxes. This transformation, in some small part, has been due to the pull from the CSPs as a direct result of the virtualization envisioned in the 2012 white paper mentioned in Chapter 1.

Let's return to economic drivers. One way to consider this is that for every bit that moves through the network there is a finite cost borne by the operator in equipment, power, and support costs. Since the introduction of the first Apple iPhone in 2007 the profile of the traffic through the network has changed dramatically, and as a direct result so have the economics of building and operating the network.

### 2.6.1 Data Traffic Dominates the Network

Figure 2.2 shows the growth in the CSP network from 2001 to 2014. This interval is chosen to clearly show the impact that the introduction of the iPhone and subsequent mobile devices has had on the usage pattern of the network. Prior to 2007 the majority of the traffic through the network was voice and the CSPs were able to monetize the use of the network by charging for voice minutes irrespective of the origination or termination on their own or another network. As long as one leg of the call was one of their subscribers, they were

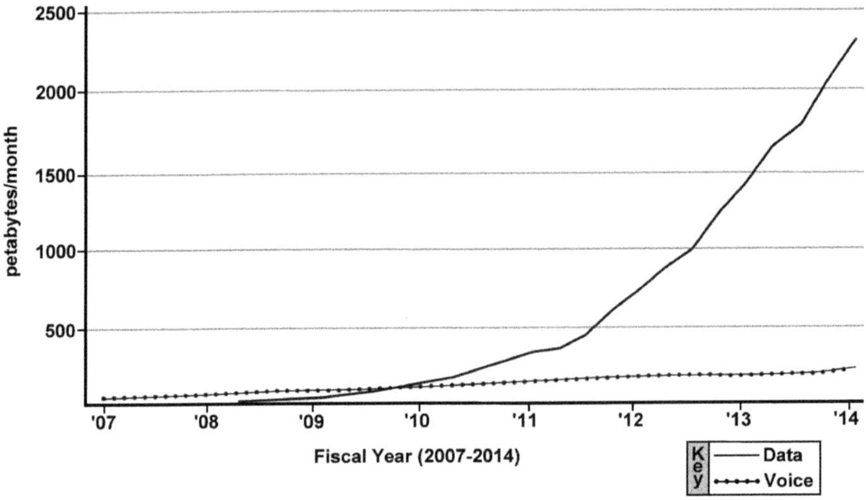


**Figure 2.2** Exponential growth of data in the CSP network from 2001 to 2014.

able to charge for the use. Just two years after the introduction of the iPhone, data usage on the network surpassed voice traffic and the race was on. Today, voice traffic through the network is barely measurable and some services carry voice as data and counted it as data traffic. Today we still call them phones, but the meaning as a sound-transmitting device is nearly lost in this context.

### 2.6.2 There Is a Fixed Cost to Moving Bits

There is a crossover point where it is estimated that the cost to build and operate a CSP network exceeds the revenue generated from the users of the network. This assessment is based on the "fixed cost to move a bit" analysis. When the cost to grow or expand the network is linear and the growth in the demand (capacity of traffic through the network) is greater than linear, there is a point where the cost to grow and operate the network crosses over the revenue curve. This would push a commercial enterprise into a losing position financially.

A very closely guarded financial analysis by some CSPs estimates that in the near future this crossover point—to expand the network capacity and to deploy new 5G nodes—may be reached. To prevent this from occurring there are only a few options: stop expanding the network, increase the fees charged to users at a rate at least equal to the demand growth rate, find new revenue sources, or change the cost curve associated with expanding and operating the network.

There is always going to be a fixed cost to move a bit through any given network and the revenue to sustain that movement of traffic must be borne by the subscribers of the network, either directly or indirectly.

It was with these realities in mind that individuals from the R&D arms of some of the leading CSPs met in 2012 to discuss how virtualization could be used to address the problem. This resulted in publication of the famous NFV white paper that was mentioned in Chapter 1.

The vision was simple to articulate but by no means easy to achieve. Having a clear understanding that the hyperscalers (cloud providers such as Microsoft, Google, Facebook, Amazon, Alibaba, and Tencent, etc.) had already faced this problem the CSPs agreed that their technologies should be leveraged. The hyperscalers had already decoupled their data center compute and networking from purpose-built systems and migrated to standard high-volume servers with virtualization. In fact, some of them had developed much of the necessary supporting software. Furthermore, in addition to transforming the cost model to build the network, they had transformed the model of providing the support and lifecycle management of their data centers. This transformation allowed them to apply a development and operations (DevOps) approach to managing the systems. This model significantly transformed the number of support engineers necessary to maintain the systems. By some estimates the hyperscalers were nearly two orders of magnitude more efficient than the CSPs in this area. Where a CSP engineer may on average maintain a few hundred systems, the hyperscalers would manage tens of thousands. In addition, the hyperscalers could upgrade software and apply patches nearly continuously, whereas a CSP may be spending months on applying a single upgrade to the systems in their network. This difference is significant if you consider the OpEx expense implications. However, and fundamentally, the regulatory model that the CSPs operate under is significantly different than that of the hyperscalers that informs how they operate and how they introduce new technology. For example, a CSP may have a significant financial penalty from a regulator or customer for failing to meet a SLA, failing to complete a call for emergency services that could lead to injury or loss of life, or billing errors that result in customers being overcharged. This in many cases drives their operational behaviors.

### 2.6.3 A Tale of Two Models

Table 2.1 contains a small sample of data from the end-of-year 2020 financial reports of two cloud and telco companies. In considering some of what may be implied in Table 2.1 note that this is not exactly an equal-to-equal comparison across companies for several reasons, one being there is no indication of the number of systems under management or the amount of power consumed to operate these systems. One nevertheless might see relationships that provide insights that can be used to argue that hyperscalers may be more efficient than CSPs for reasons worthwhile to explore.

**Table 2.1**
Financials of Four Representative Companies from 2020

| | Revenue (B USD) | Market Cap (B USD) | Employees (k) | CapEx (B USD | OpEx (B USD) | P/E (B USD) | EPS (USD) |
|---|---|---|---|---|---|---|---|
| **AT&T** | 171.7 | 205.8 | 230 | 15.7 | 165.4 | 0 | –7.5 |
| **Verizon** | 128.3 | 240.7 | 132.2 | 18.2 | 99.5 | 13.22 | 4.3 |
| **Google** | 182.5 | 1916.8 | 135.3 | 22.3 | 141.3 | 29.9 | 58.61 |
| **Facebook** | 85.9 | 1060.4 | 58.6 | 15.1 | 53.3 | 27.07 | 10.09 |

This data was accumulated from the annual reports (10k) from [2–5].

It is readily acknowledged that this table contains a mix of companies, and that this is only one small sample, and that even among the two communication service providers, their business models contain different segments that may impact the analysis. The intent here is not to provide detailed analysis of the balance sheets of these companies; that can be left to the business school authors. The intent is to provide some backup to the claim that the transformation of the network that generates the revenue, and therefore profit of the CSP, is challenged. It would be interesting to consider a similar analysis for the CSPs on a global scale and analyzing their changes over some period of time; for example, to consider the change in headcount each year since 2010. Headcount of course is a leading indicator of the staffing efficacy to maintain and operate any given network.

## 2.7 Applying the Cloud Model to the Telco

The observation by the leading CSPs from the 2012 white paper should then be clear: as much as possible, separate the hardware from the software using modern techniques (at that time virtual machines (VMs)), also known as disaggregating the hardware from the software, implying it may be purchased from different vendors. Figure 2.3 captures this concept of disaggregation. The NFVi layer is a standard high-volume server and the virtualization infrastructure middleware (e.g., for Type 2 hypervisors a host operating system, a hypervisor, and a guest operating system). The details of this will be covered in Chapter 3. This model predated the current trend to use containers, which will also be discussed in Chapter 3. The VNF 1, VNF 2, and so on, represent the interesting workload running in the network and this is where the term VNF was coined. It is not uncommon for someone new to this field to easily confuse this with NFV, which is a generic description for the overall concept of disaggregation of software and hardware, rather than a specific instance of an interesting workload. A simple example of an interesting workload might be a firewall running in a

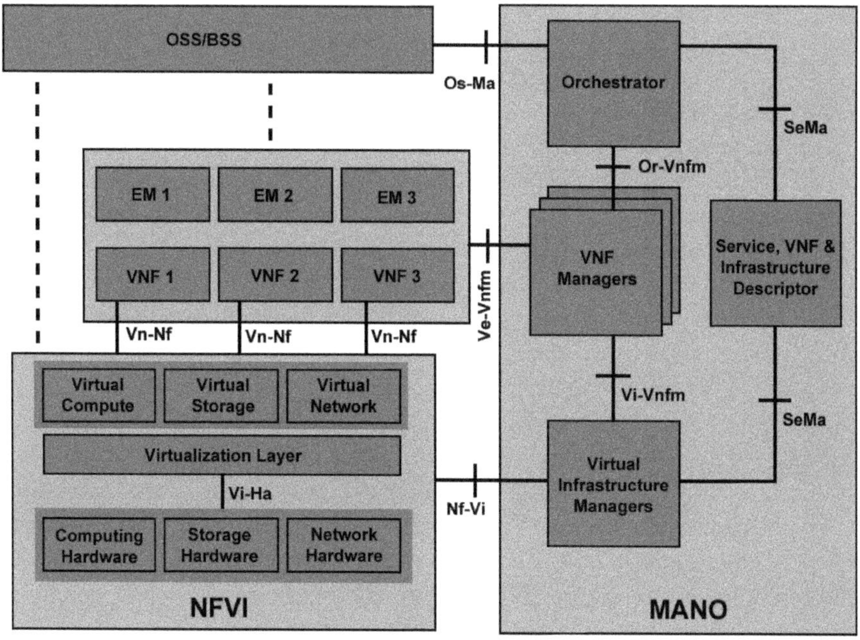


**Figure 2.3** The NFV architecture.

VM. Typically, the VMs running above the NFVi handle the traffic (control and data) passing through the network. One interesting element in Figure 2.3 is the EMS. These vendor-specific functions are not directly involved in the traffic flow; rather, they provide a system control point for managing events, alarms, and general control of the running workload. The EMS function today may be becoming less and less incorporated in the NFV framework depending on the vendor and their EM approach.

To the right in Figure 2.3, NFV or MANO is not directly involved in the traffic flow; rather it is acting in a supervisory role. Again, this will be discussed in more detail in later chapters.

NFV changes the procurement model to purchasing and installing hardware on a hardware schedule and licensing and upgrading software on a software schedule, thereby moving at a speed more typical of the hyperscalers, albeit with the modifications needed to align with the CSP business model such as providing lifeline services and compliance with regulatory mandates.

Marc Andreessen has been quoted that "software is eating the world" [6–8]. Yet hardware won't just go away, after all, it's not turtles all the way down [9]. The significance of vendors providing solutions into the telco industry where historically they have led with hardware, pivoting to one where the application or solution provided by software, is the leading factor challenging the CSP operational network today.

## 2.8 Paths Taken to Evolve the Telco Network

The emergence of the cloud started around the turn of the millennium and grew rapidly in their first two decades. The first iPhone sparked the emergence and subsequent rapid growth of what many call over-the-top (OTT) services in the wireless network. Some of these services were running on hyperscaler infrastructures and nearly all were in the cloud. This generated a forcing function within the telco network as now a significant amount of traffic was data that simply flowed through the network without the network adding much value, if any. Up until this point the telco's primary source of revenue from their mobile network came from a combination of voice calls and short message service (SMS) text messages. Each was priced separately in many cases at the turn of 2000. The average revenue per user (ARPU) soon included a significant volume of data as part of a monthly allocation or as part of a prepaid plan. Data connectivity through the mobile network at the introduction of the first iPhone [10] was significantly limited. This was prior to the capabilities introduced with the 3G [11] network rollout (the iPhone came before 3G was fully enabled and the first iPhones ran on late 2G network technology initially), the maximum throughput being somewhere between 120 to 384 Kbps.

### 2.8.1 3G Data Begins to Be the Primary Content in the Network

The introduction of 3G brought two significant changes into the global network. The first was a truly global standard for the radio interface at both the user equipment and the access point. Converging on a common global standard has cost benefits at scale as vendors can design and manufacture devices for the global market. This allowed for consolidation of technologies at global scale. There was truly one global standard. Handsets (called user equipments in the industry and standards) and radios were standardized; you no longer had one set of standards operating in Europe and Asia and a different set of standards operating North America. This unification allowed for a broader base on which amortization of the development costs could be spread. While we may choose to baseline the network transformation with the introduction of the iPhone and the global deployment of the 3G network, there were also other forces at work.

### 2.8.2 Interfaces Connecting Endpoints in the Network

Until 2007 much of the traffic that flowed over the mobile network was still primarily voice calls. These calls took a different path in the 3G era (until Voice over Long Term Evolution (VoLTE) in 4G) within the computing nodes in the network than did the SMS (and soon MMS) and data traffic. We know that people still talk, and that using the mobile network for voice does not require everyone to be on the same "network"; two people can be subscribers of different network providers and don't care which network each other is on. The communication service providers (with the assistance of a global organization that will be discussed later in this chapter) long ago sorted out the interoperation of their traffic, providing paths from one network operator to another, both regionally, nationally, and internationally.

The digitization of the voice traffic through the network took place decades ago. Voice calls through the network consist of two components, often called the signaling or control plane, and the data plane carrying the actual voice traffic. The complete separation of these two elements, control plane and data plane, have progresses slowly. The network until recently was primally built on what today is called purpose-built equipment, and the ecosystem providing these systems consisted of about 12 very large-scale vendors, many of whom have worked diligently to protect their legacy proprietary product lines.

## 2.9 The Ever-Evolving Introduction of Technology into the Network

Contrary to what might be perceived as a static business, the telcos have been evolving for a very long time, and in fact were initial innovators in many areas of technology. There are several ways to view the start of the transformation of any legacy business and most involve a process of evolution over revolution. The CSPs have done it before and there is no practical reason to believe that they are not going to be successful this time, even in the presence of internal antibodies that resist change.

Looking back to the last millennium, the protocols and physical interfaces that made up the global CSP network were based on global standards to ensure interoperability. Standards are expensive to develop and expensive to certify. Often, equipment (hardware and the accompanying software) has to be certified for each network.

### 2.9.1 Making the Network Global

Starting in the mid-1950s, telcos began consolidating into a global standards development effort under what was known as the CCITT (abbreviated in French for International Telegraph and Telephone Consultative Committee). The CCITT was organized as a specialized agency within the United Nations and was founded much earlier, in the mid-1800s. Today, the organization is known as the ITU-T and is responsible for the development of international standards critical to the information and communication technologies (ICTs) [12].

One of the efforts was the development of a global standard for the signaling protocols that evolved into what was called Signaling System 7 [13, 14].

This system of call and session control evolved quickly in the 1980s and 1990s as the mobile network was being developed and deployed globally.

There were four large geographical standards with slight differences in the protocols that were developed. These were North America led by the American National Standards Institute (ANSI), Europe led by the European Telecommunications Standards Institute (ETSI), and two leading standards in Asia, one led by Japan, and one led by China. Some consider that these four standards were developed to encourage a local ecosystem of vendors and to manage competition due to the cost of implementation and compliance testing, while others consider that the motivations are primarily technical (e.g., "we don't need 14 bits for that, 12 is enough"). These arguments today feel like editorial compromises since most implementations were on octet aligned structures.

### 2.9.2 This Global Network Comes at a High Cost

It was not uncommon for early signaling nodes with only a few low-capacity interfaces to cost multiple hundreds of thousands of U.S. dollars for highly reliable computers. Today, in the Internet Protocol (IP) space, these would be simple servers with a few Ethernet ports costing a few thousand dollars.

This was the time when the first-generation mobile network was being developed. This initial technology was based on adjuncts being added to the existing telco systems. Endpoint class 5 switches were supplemented with dedicated hardware that allowed mobile phone calls and for the usage to be appropriately billed.

Around the same time (the early 1980s), Ethernet and the IPs were evolving rapidly and being deployed in enterprise IP networks. These systems initially were operating at a data rate of 1 Mbps, a rate similar to the T1/E1 rate found in existing telco equipment at the time, and yet the cost of the interfaces and the associated cost for the supporting software were orders of magnitude less than the cost of the telco network interfaces. Of course, there were significant differences; one was in the ability of the IP to load balance and provide redundancy that was core to the telco network. Early work using IP to build resilient systems did begin shortly after deregulation of the U.S. market in the mid-1980s with the introduction of service nodes for control signaling functions.

The protocols were different and the traffic that was typically carried over these protocols at that time was also significantly different. While the IP model was founded on packet switching, the telco model was circuit switched. The Internet Engineering Task Force (IETF), the primary standards body around all things related to the IP, was developing Request for Comments ((RFCs), their equivalent to a standards document) based on generally working code. RFCs were progressing at a pace only dreamed of by the telco standards bodies. In addition, there were no national or geographical differences in protocols, leading to a better economic model [15].

A simple way of viewing this is shown in Figure 2.4. Today, no one would intentionally consider deploying new T1/E1 capable equipment if there was any other option (there may be some remote interworking needs where carrier-to-carrier interfaces are still being deployed using legacy TDM interfaces).

Before moving on it is worth noting that many consider the value of the network to be based on Metcalf's law [16], which states that the value of a telecommunications network is proportional to the square of the number of connected users of the system. This consideration may also be seen in the value placed on social networks [17]. This model has not made an impact on the market value of the public telcos and while there is good alignment on the concept, we would suggest that value in this case may have a different scope than the market forces placing a valuation on any specific telco.

### 2.9.3 Relating This Back to the 5G Network

All this discussion on protocols and standards is significant as the evolution of the mobile network from 1G to 5G and beyond was built with support for backward compatibility, and where possible, common infrastructure. Early cell towers were connected back to central offices via T1/E1 interfaces in many cases. These interfaces were on twisted-pair copper that had limited capacity but were well understood by the operators who were concerned about reliability in the field.

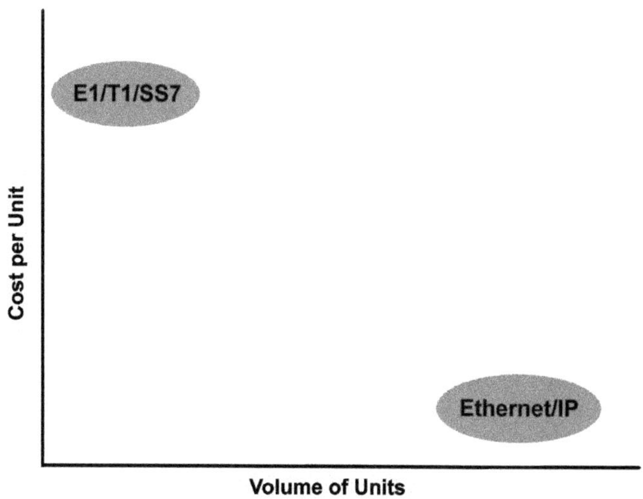


**Figure 2.4** Number of interface units versus cost.

The alternative was (and in some geographies still is) to use microwave relay for point-to-point connection from a tower back to a central office location. The choice to use microwave impacts both the cost structure (initial and operational) and maintenance required to keep the system operational. Today, a significant portion of cell towers are connected via long-haul fiber and in many cases Ethernet protocols have replaced the legacy telco protocols. All this was done while the network remained fully operational as experienced by the user.

## 2.10 The Drive for Improved Agility and Efficiency

Agility and efficiency are two qualitative goals to the disaggregation that span both the CapEx and OpEx challenges facing the CSP. Let's break down the CapEx first for both, then deal with the OpEx.

Agility in the CapEx model is aimed at liberating the operator from what is often considered vendor lock-in and in addition the long deployment and upgrade cycles, bringing nimbleness into the process of building the network. Ideally, a CSP will procure a compute resource, to their specification, from a number of approved vendors, have these systems installed and operationalized in the network in cloudlike time frames. Traditional planning cycles at the CSPs have had multiple years of lead time, procurement processes taking up to a year and many months until utilization in the production network. The CSPs realized that the increasing traffic load and the variability of the traffic their network was exposed to could not sustain this legacy model and the cloud or hyperscaler enterprises had already successfully navigated these waters. For example, a CSP might have a normal cadence for adding standard high-volume servers into their mobile core in anticipation of capacity growth. These servers could follow a release cadence from a hardware original equipment manufacturer (OEM) or original design manufacturer (ODM) where about every 12 to 18 months new servers with improved capacity and performance are available at close to the cost of the previous generation of servers.

### 2.10.1 DevOps and Continuous Integration and Continuous Delivery

DevOps [18] is a concept that was first found in the IT industry, where the development and operations teams and their workflow were unified in a single practice for operationalizing software. The practice was driven by the need to move faster from development to deployment in operational networks. In many cases this was driven to improve quality, reduce delays in bringing new features to market, and fixing defects faster. The natural extension of DevOps is to implement a model of continuous integration and continuous delivery (CI/CD) culture. This efficiency and agility work well in the cloud environments and the telcos have a great desire to adopt these practices in their production network.

The telcos have one significant difference in the model found in IT: they in general do not produce the software that is functional in their network and instead rely on their TEM vendors for their operational software. Ericsson has published several papers discussing telco DevOps where they lay out a vision of adapting DevOps into the telco network [19].

DevOps in the telcos has several challenges; one is that the critical operations of the national assets of the network resist introduction of unproven software, and in some cases the existing systems do not have the ability to partition test traffic in the live network. This may prevent unproven software from being deployed until the carrier has performed extensive testing to validate the new release of software. Part of this is cultural, and part is technical. Both these problems require resolution to fully achieve the agility found in cloud-native DevOps. One example of the technical challenge is allocating a new release of software to specific nodes in the network and to only allow test traffic to be routed to these nodes. The test traffic would not be capable of fully loading the systems in many cases and while capable of proving limited feature traffic validation, may also lack the full spectrum of variability found in the live traffic. This remains an active area of work and is likely to continue for some time in the production network. The IEEE has published a document capturing the requirements and guidance on the implementation of DevOps [20].

## 2.11 Separation between Data Plane and Control Plane

The control and data planes have been heading toward separation for a very long time in the network. One term often used today is control user plane separation (CUPS). For those unfamiliar with the terminology of control and data, a simple example may be useful here. The control plane carries the signaling and other associated messaging necessary for the establishment of any communication and sessions. Breaking this down into what may be a familiar description on how to establish a voice connection from one person on a UE to another person on their UE (even though the number of voice calls is rapidly shrinking it still serves to show the separation of the planes). If communication from one UE to another is on the same network, the originating UE offers an invite of some type into the network and the network acknowledges the invitation and determines the path to send the offer to the destination UE (if available). The destination UE accepts the invitation to establish the call by sending a control message back toward the originating UE which would then acknowledge that the request is still valid. The network will then establish a full duplex data plane session between the two endpoints, allowing the human voice to flow from end to end.

It is usual to place all the control messaging and associated sizing onto servers separate from those responsible for the transfer of the voice data (in RTP messages). RTP is the Real Time Protocol, a specification for the conversion of voice into packetized data for transmission over a digital interface. Figure 2.5 shows the functional elements at a high level in the 5G Core network. The AMF receives the control messaging and associated session management messages from the UE. Control for the connection and mobility is maintained by the AMF and any session messages are sent to the session management function (SMF) (a voice call, for example, is a "session"; there can also be other types of sessions). The AMF and SMF may in fact reside on the same server or cluster of servers, or in some cases may also be realized on different servers as a design or scale decision of the manufacturer. What is important is that the functions adhere to the specified interfaces; in this case the AMF and SMF communicate in accordance with the N11 interface.

We will have more to say about the other nodes (AUSF, UDM, PCF, and AF on the control plane) in later chapters.

### 2.11.1 The 5G User Plane Function and Data Network

The two elements shown at the bottom of Figure 2.5, the UPF and DN, are the data plane elements of the 5G Core.

The data network (DN) is the service provider path or connection to the internet or other third-party services that may be resident in their network or off their network. The DN module shown in Figure 2.5 may in fact have multiple destinations and even multiple instances in the same network, all adhering to the N6 interface.

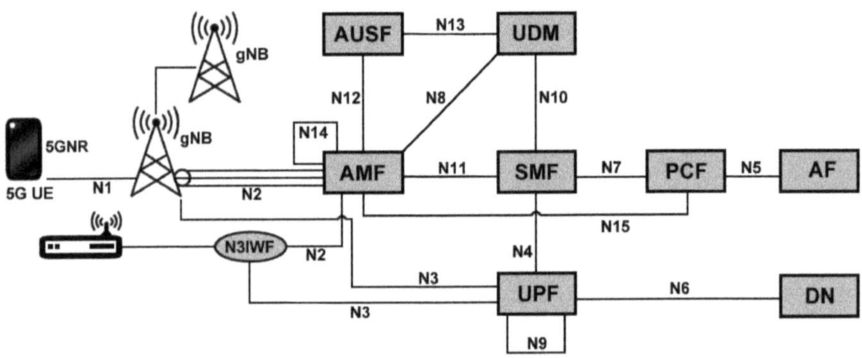


**Figure 2.5** The 5G Core network nodes.

More interesting in terms of 5G Core is the user plane function (UPF). This node is the anchor point for the mobile device for both inter- and intra-operations (within one or multiple networks). The functions provided include the external protocol data unit (PDU) session interconnect point to the data network, packet forward and routing, and packet inspection. The user plane function part also provides policy rule enforcement, if any (e.g., redirection and traffic steering), and, if required, any lawful intercept of the user part data. Additional functions include traffic usage reporting, quality of service (QoS) enforcing, uplink and downlink (UL and DL) rate enforcement, and QoS marking in the DL, as well as UL traffic verification in QoS flow mapping. Transport-level packet marking is included for both the UL and DL by the UPF node in addition to DL packet buffering and DL data notification triggering, and finally, sending and forwarding one or more end markers to the 5G RAN. One would not be expected to recall all these functions found in the UPF, but one should be expected to understand that this is not simply a router function with multiple ingress and egress ports. One function, the requirement for lawful intercept (LI) is of interest due to the possibility that different jurisdictions may place different requirements on this function to protect the knowledge that certain users are being monitored for possible illegal activity. We bring up LI because while the standards are global, this is one possible example where a global standard may not suit all local requirements.

The design goals of 5G have nearly completed the separation of the control plane from the user plane; nearly, because there has to be some direct interaction between the two. The control plane components are still in communication with the data plane components for what should be obvious reasons. The separation actually allows two key goals to be reached: first, the scaling of each plane is now independent of the other plane, and second, the ability for them to be placed at different locations if and when required or desired.

An additional benefit may be realized when systems are used to support multiple protocols at the same time, such as might be found when both 4G and 5G are used in the same network. Reuse of infrastructure may also provide for slicing of transport networks and scaling of control and management dependent on the transient nature of the traffic in the network at any given time.

### 2.11.2 5G Standalone and Non-Standalone Deployments

Introduction of 5G into the network follows a long history of new versions being introduced while previous generations are still operational. Early 5G deployments saw the RAN being deployed on the new 5G Radio or Next Generation (NG) radio deployed while the core remained unchanged; these deployments are called Non-Standalone (NSA). In these deployments the signaling portion remains compliant to the 4G specification while the RAN interfaces are 5G-compliant. The second model is Standalone (SA); in this model the RAN and core are all operating at the full 5G standard for both the control and data plane. While neither is a better choice, some telcos found the NSA a faster route to market with 5G and others may have chosen SA, in particular in locales where there was little or no legacy 4G. In many cases this came down to a financial decision: roll out the 5G New Radio and continue to operate on the 4G core until a later budget cycle.

## 2.12 3GPP as the Leading Standard Body for the Mobile Network

The CSP mobile network has a very long legacy of standardization, and even prior to the first mobile phone network the telecom network was well described by standards bodies. As mentioned earlier, during the time of 2G it was well recognized that a global standard was required to drive costs lower and increase interoperability across geographical boundaries. This resulted in the establishment of what is today considered to be the most significant standards body in the mobile network: the 3rd Generation Partnership Project (3GPP) that unites seven organizational partners [21]:

- The Association of Radio Industries and Businesses (ARIB), Japanese standards body;
- The Alliance for Telecommunications Industry Solutions (ATIS), North America;
- The China Communications Standards Association (CCSA), Chinese standards body;
- ETSI, European standards;
- The Telecommunications Standards Development Society, India (TSDSI), Indian standards;
- The Telecommunications Technology Association (TTA), Korean standards;
- The Telecommunication Technology Committee (TTC), Japan.

While the name retains the 3G label, the body has moved on to establish the 4G, 5G, and soon the Beyond 5G (which naturally, some are already calling 6G) standards.

The work of this body is massive, although a new technical entrant in the field would be well served by spending some time becoming familiar with the work captured by the 3GPP. A good starting point might be the very high quality "About" page on the 3GPP portal [22].

## 2.13 Introducing the International Telecommunication Union

While 3GPP is an industry standards body, the International Telecommunication Union (ITU) is another significant standards body that needs to be discussed. The ITU is an agency of the United Nations with a history that goes back to 1865. Today, the ITU is divided into three sectors known as ITU-D, ITU-T, and ITU-R. One of the most significant responsibilities filled by the ITU for the mobile network is establishing global standardization and allocation of spectrum. There are several reasons this responsibility lies with this special agency of the United Nations. Fundamentally for ITU-R it comes down to management of spectrum allocation. Spectrum, a finite physical resource, knows no national boundaries, and while some jurisdictions are large and isolated—for example Australia—there are many others where this is not the case. Even though spectrum is global, each nation has the rights to the usage, and it is a significant monetization opportunity for governments through licensing. The allocation of spectrum allows for global interoperation with an economy of scale for the manufacturer of radio devices. The regulation also provides for controlled management of any interference through the allocation of guard bands to prevent one radio channel band from interfering with another band. Further adding to the complexity is the business consideration that nearly every jurisdiction has multiple operators providing cellular service. Each of these operators needs noninterfering channels to provide connectivity to their users when operating in the same area as another cellular operator. In addition, the use of spectrum for satellite communications must be regulated and since satellites span national boundaries the allocation of spectrum to them are under the regulatory control of the ITU-R. One additional area of concern (there are likely others too that may emerge over time) that has arose with 5G involves the navigational systems of commercial aircraft that operate in the 4-GHz range. To date, this concern has not materialized into any known impact but remains an active area of investigation.

The ITU operates on a four-year cycle of study and standardization when, at the conclusion, nation states ratify via treaty the adoption of the recommendations of the ITU. Spectrum allocation to new bands for the mobile network begin with work in 3GPP, are studied by members of the ITU-R to ensure usage does not break existing band usage for other allocated functions globally, and finally after ITU-R ratification the 3GPP will adopt the standards and the investment to implement the radio by vendors moves forward in earnest. This is one very significant reason that the progress of each generation takes nearly a decade.

Today the ITU standards are available online. Long gone are the days of the ITU publishing their standards in hard copy and charging a hefty fee to just obtain a copy [23].

The ITU still has a significant role in ensuring international connectivity between operators and nations in the communication network. Fundamentally, the ITU-T enables the global ability to make phone calls to any end point on the globe, access the internet (national firewalls being taken into consideration) from any place on the globe with mobile access, and in many cases from landline access.

## 2.14 Other Standards Bodies

There are other standards bodies that contribute to the telco space in various ways. Some of the most significant include the IETF, ETSI, and the Institute of Electrical and Electronics Engineers (IEEE) [24].

The IETF should be well known by many, if not directly then certainly indirectly. This open international community is comprised of researchers, designers, operators, and vendors working to architect and ensure smooth operation of the internet. This body published their work in RFC as standards. Many of the protocols today found in the core of the network are the result of the open work by the many members of this body.

The IEEE is the world's largest professional society. Their work spans a significant range of technologies, several of which find their way into the communication network, and many also which do not. The IEEE may be best known as the body responsible for the Wi-Fi standards. The IEEE Communication Society (ComSoc) [25] has a number of conferences and publications that share the work of forward-looking researchers in the field of telecommunications.

ETSI, based in Sophia-Antipolis, France, was founded in 1988 and has nearly 900 members from at least 61 countries on five continents. ETSI is one of only three standards bodies officially recognized by the European Union as a European Standards Organization. ETSI standards are the only ones that can be recognized as European Standards. As a result, many CSPs in Europe (and also in other jurisdictions) mandate compliance to the ETSI standards in the equipment installed in the operational network. ETSI is extremely active in the standardization of 5G (and going back to 3G) and will be a significant contributor to the future 6G standards.

The TM Forum [26] is yet another body with over 850 members in the telecom field. This community works to help the service providers and their vendor community in delivering the digital transformation by providing an open and collaborative environment along with practical expertise in transforming the business systems and operations. This is often referred to as the operational support systems (OSS) and business support systems (BSS) [27].

The Open Networking Foundation (ONF) [28] is an operator-led nonprofit consortium working to transform the network infrastructure and business models of the telecom industry. Their efforts are in the areas of advocacy, education, development, and applied research. They trace their origination to 2011 with the emergence of SDN and the resulting work on the OpenFlow protocol. The leading partners include some of the world's leading CommSPs, including AT&T, China Unicom, and NTT Group, and industry leaders Tech Mahindra, Google, and Intel. There are in addition over 100 member partners contributing to the consortium.

One additional important body is the Linux Foundation (LF). While not technically a standards organization, the work performed by LF participants is a significant contribution to the NFV transformation. The Linux kernel and many supporting projects are hosted by LF; with over 800 open-source projects there is a wealth of opportunity to both learn from existing efforts and contribute meaningfully to the community.

## 2.15 Open RAN's Role in Virtualizing 5G

The O-RAN Alliance [29], also known as Open RAN, is one of the leaders in enabling the transformation of the radio access network (RAN) today. Open RAN is a concept, whereas the O-RAN Alliance is an industry body working to specify an instance of an Open RAN. The only Open RAN today is being specified by the O-RAN Alliance and the terms are often used interchangeably as there is currently only one specification in development, from the O-RAN Alliance, for an Open RAN. This operator-defined alliance is driving (some may say pulling) the industry toward an open, virtualized, and intelligent RAN. With 30 operator members and nearly 300 members from industry and academia, their scope is far reaching and global.

Since the founding of O-RAN in 2018 the impact on the industry can be gauged by the growth of the ORAN ecosystem and the creation of architectures as the foundation for an open, intelligent, and a virtualized RAN. This is accomplished through collaborative standards development, ecosystem testing in global plug fests, and integration efforts. O-RAN has published open software that can be found in [30].

In addition to disaggregation of software from hardware in the RAN, ORAN is prescribing open hardware architecture and virtualization. Their goal is to specify a high performing and scalable control layer and truly open interfaces between all the RAN nodes through standards that are well defined, thereby enabling the operator to use a multivendor supply chain. The work includes enabling ML for control of the RAN resources to optimize performance and services. This work complements the work of 3GPP and other industry standard bodies and is not a duplication or replacement of those efforts. The goal is to enable a broader industry ecosystem and foster faster innovation and adoption into the operator's networks.

There is one topic that arises in discussions at times that can be a source of potential confusion when discussion O-RAN. In the abstract, an open RAN need not be virtualized, and a virtualized RAN need not be open. Fortunately, these two abstract concepts are in different technical dimensions; that is, they can coexist without conflicting with each other. In fact, the O-RAN Alliance, while having "open" in their name, is working on the disaggregation, virtualization, and open needs of the community. The work of O-RAN will be covered in more detail in Chapter 6.

## 2.16 Venture Capital Investments

It has been nearly two decades since there was significant venture capital (VC) investment being made in the telco market. There may be a number of reasons that the VCs are not as active in this field today. One is the very high costs of entry as evidenced in the above discussion; the volume of work necessary just to enter as a vendor and provide backward compatibility takes significant time and investment. In addition, bringing new innovations to a mature market requires the procurement organizations to be able to introduce the new technology or vendor. This is a significant deviation from the innovation that can and does occur in the cloud. Consider the effort to establish a service like Airbnb or Uber. There is no legacy infrastructure to support, no existing procurement rules to comply with, no compliance testing to pass, no international standards to meet; rather a great business idea and the right software to implement the business model. There are some pockets of telco innovation; these usually come in the form of the ISVs. Prior to the internet bubble burst of March 2000 there was significant VC activity around the telecom industry. A number of the most innovative companies were acquired prior to the burst and as mentioned earlier Cisco was an acquisition leader in the field. Others simply did not survive due to the economics of time more so than a failure in the technology. There are some who believe this experience remains one of the legacy reasons that VCs remain cautious today. Two companies have recently demonstrated that the model of early VC investment followed by acquisition by a larger tech company is still valid. Microsoft acquired Affirmed Networks in March 2020 and then Metaswitch in May 2020 [31, 32]. These two ISVs provided Microsoft with the technology to enter the core of the mobile network, and today we are seeing Microsoft apply this technology in Edge deployments where they may soon be competing with the traditional CSPs for core services to mobile and possibly fixed-line subscribers both in the consumer and enterprise spaces.

Rakuten, as a 5G green-field operator in Japan, has developed a network from the ground up with a goal of virtualizing the entire network. They are the most recent example of disruption in the market and continue to innovate their business model. One way is to offer their designs and technology to others building new networks or looking to transform from the traditional NEP-provided network models. A recent movement was the acquisition of Altiostar [33, 34] in August 2021. Altiostar had been a supplier of RAN components into Rakuten [35]. Rakuten continues to innovate on the business front to reduce their network rollout costs.

There are other leading ISVs that are emerging; one is Mavenir [36]. Mavenir, while not as young as some of the other ISVs, has been developing a software-first offering (sometimes called a pure-play software) for many years, and in the past decade has evolved into a significant disrupter in the 4G and today in the 5G RAN and mobile core space. Mavenir has been a significant force in the O-RAN work and continues to innovate, placing pressure on the more traditional RAN vendors.

## 2.17 Summary

We have attempted to establish the business motivation behind this latest transformation, transforming the cost model to build and operate the global mobile network. This is driven by the ever-increasing volume of data that the network has to transport, along with customer resistance to increases in their average monthly bill irrespective of data usage. ARPU is a metric by which mobile operators report their financials, and it is a legacy model that is unsustainable. Virtualization, more precisely disaggregation of the nodes in the network that comprise the networking, compute, and storage, enables a telco to better negotiate the cost of hardware and software that comprise their network components. This trend is challenging for many, in both the ecosystem and within the telcos themselves.

This chapter also spent considerable time discussing the various standards bodies engaged in this industry. This is seen as critical for any enterprise choosing to enter or remain in this field because they will have to participate in, be aware of, and possibly comply with, a number of international standards and also fully understand the implementation details of the standards. One common question is: Standards or open? In our view this is an inappropriate mixing of ideas; one can have both standards and open as can be seen with a number of the bodies mentioned above, in particular the work of the O-RAN.

## References


[1] https://techterms.com/definition/greenfield, "Greenfield," November 2022.

[2] https://investors.att.com/~/media/Files/A/ATT-IR-V2/financial-reports/annual-reports/2020/complete-2020-annual-report.pdf.

[3] https://www.verizon.com/about/sites/default/files/2020-Annual-Report-on-Form-10-K.PDF.

[4] https://abc.xyz/investor/static/pdf/20210203_alphabet_10K.pdf?cache=b44182d.

[5] https://investor.fb.com/investor-news/press-release-details/2021/Facebook-Reports-Fourth-Quarter-and-Full-Year-2020-Results/default.aspx.

[6] https://quoteinvestigator.com/2018/01/24/software.

[7] https://genius.com/Marc-andreessen-why-software-is-eating-the-world-annotated.

[8] https://www.forbes.com/sites/peterbendorsamuel/2019/10/14/software-is-eating-the-world-but-services-is-eating-software/?sh=1743c0182116.

[9] https://en.wikiquote.org/wiki/Turtles_all_the_way_down.

[10] https://www.apple.com/newsroom/2007/01/09Apple-Reinvents-the-Phone-with-iPhone/.

[11] https://en.wikipedia.org/wiki/3G.

[12] https://www.itu.int/en/ITU-T/about/Pages/default.aspx.

[13] https://en.wikipedia.org/wiki/Signalling_System_No._7.

[14] https://www.itu.int/rec/T-REC-Q.700.

[15] https://www.ietf.org/.

[16] https://spectrum.ieee.org/metcalfes-law-is-wrong.

[17] https://www.peterfisk.com/2020/02/metcalfes-law-explains-how-the-value-of-networks-grow-exponentially-there-are-5-types-of-network-effects/.

[18] https://devops.com/what-is-devops/.

[19] https://www.ericsson.com/en/reports-and-papers/ericsson-technology-review/articles/devops-fueling-the-evolution-toward-5g-networks.

[20] https://ieeexplore.ieee.org/document/9415476.

[21] https://www.3gpp.org.

[22] https://www.3gpp.org/About-3GPP.

[23] https://www.itu.int/en/history/Pages/AssemblyTelegraphTelephoneTelecommunication.aspx?conf=4.260.

[24] https://www.ietf.org/.

[25] https://www.comsoc.org.

[26] https://www.tmforum.org.

[27] https://www.forest-interactive.com/blog/what-are-oss-and-bss.

[28] https://opennetworking.org.

[29] https://open-ran.org/.

[30] https://wiki.o-ran-sc.org/display/REL/Releases.

[31] https://www.affirmednetworks.com/microsoft-has-signed-a-definitive-agreement-to-acquire-affirmed-networks/.

[32] https://www.metaswitch.com/blog/metaswitch-announcement.

[33] https://www.mobileworldlive.com/featured-content/top-three/rakuten-buys-jtower-stake-to-cut-network-rollout-costs?ID=a6g1r000000zRzaAAE&JobID=927145&utm_source=sfmc&utm_medium=email&utm_campaign=MWL_20211018&utm_content=https%3a%2f%2fwww.mobileworldlive.com%2ffeatured-content%2ftop-three%2frakuten-buys-jtower-stake-to-cut-network-rollout-costs.

[34] https://www.altiostar.com/rakuten-group-to-acquire-mobile-industry-innovator-altiostar/.

[35] https://www.rakuten.com/?landing&src=msn&eeid=17881&utm_channel=sem&utm_medium=sem&utm_source=361532245&utm_campaign=b&utm_content=c&utm_term=bng&utm_pub=1181975887977305&utm_size=kwd-73873619528912:loc-190&acct=b&ds_kids=p57372172605&dest_stack=act&&msclkid=c831a908f7d511815d63c87e347dc180&gclid=c831a908f7d511815d63c87e347dc180&gclsrc=3p.ds, October 2021.

[36] https://www.mavenir.com/.

# 3

# Virtualization Concepts for Networks

## 3.1 The Virtualization of the Network

Hardware virtualization is a key technology in all instances of today's and very likely all future communication networks. As a result, it is necessary to introduce the fundamentals of computer system virtualization and the basic data structures for virtualizing networks. The chapter begins with a very basic discussion of what virtual systems and computer system virtualization are. It will then cover the fundamentals and technical concepts of computer virtualization, and show the association with cloud computing and network virtualization. This foundation is necessary as the concepts and implementations developed further in this work presume a deep understanding of the foundation of virtualization. The discussion of the concepts begins with an overview of the historic development of virtualization techniques.

Virtualization contains its own concepts and terms and the progression from compute virtualization to network virtualization influenced each other. Key achievements in computer virtualization triggered subsequent developments in network virtualization and vice versa. This chapter, as well as the book at large, builds on the understanding that the paradigm that the virtualized 5G and beyond is part of the virtualization of networks in general and is not a single and homogenous subject. The concept of network virtualization is rather a combination of three main fields of virtualization technologies, each representing a stream of innovations. This chapter assumes therefore, that there are three major innovation areas that contributed to virtualized networks as shown in Figure 3.1: computer virtualization, compute infrastructures (i.e., mainly

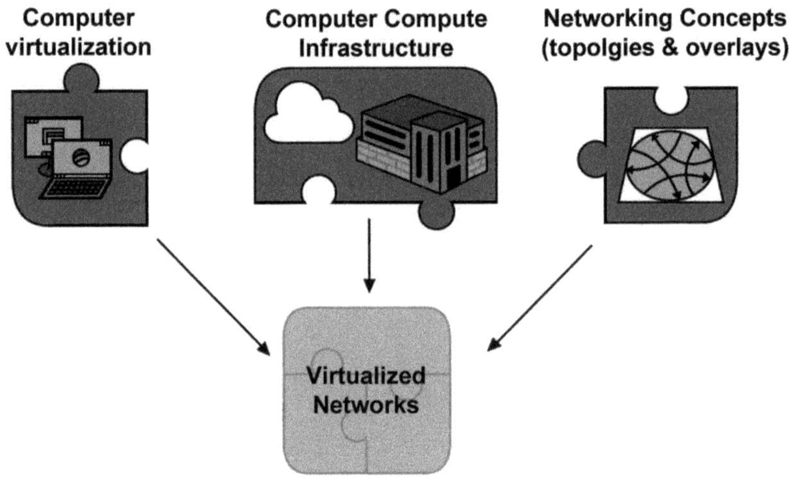


**Figure 3.1** Components leading to virtualized networks.

concepts for cloud computing), and networking concepts (topologies, overlays, networking algorithms, and data structures). This chapter provides an overview and introduction to virtualization concepts, and then Chapter 4 will detail the state of the art of the technologies for virtualizing the data plane of mobile networks.

### 3.1.1 What Is Virtualization?

The virtualization of computer hardware was considered early on to be a useful concept to enable access and to share computing resources. The term "virtual" means in general that "...something is so nearly true that for most purposes it can be regarded as true...." [1]. Applying this definition to the subject at hand, a virtual computer permits users to use a computer without knowing if the computer is a self-contained physical system (currently sometimes called "bare-metal") or a portion or subset of a machine that appears to be a self-contained physical system. Computer virtualization initially aimed at making computation a commodity (i.e., providing widespread access for conducting complex and sophisticated calculations).

A virtual machine (VM) [2] is a near true copy of its physical counterpart. A VM consists of abstracted versions of the hardware resources of a typical von Neuman computer architecture (i.e., of a virtual CPU, virtual memory, and a virtualized IO system that can efficiently be used). A VM runs its own version of an operating system, and the functions are separated from the other VMs. A virtual machine is typically said to run as a "guest" on a physical host.

Virtual computing is obtained in general by two modes when focusing on the efficiency of resource management. The first mode is denoted as the sharing mode. In this mode, a sole computing system appears as different computers. The hardware resources are split among each of those different virtual computers. If a resource is atomic; that is, it doesn't permit concurrent use (e.g., a single-instruction CPU or IO), then a sequentialization (i.e., conversation of concurrent requests following an order) is enforced.

The second mode is called the aggregation mode. Here, many physical computers appear to be a single computing system. The virtualization system bundles the incoming usage requests and redistributes them to the multiple physical resources. The aggregation mode enables a virtual computer to achieve very high performance. It provides for parallelization of requests (i.e., concurrent execution of requests). Thus, this mode increases the throughput observed by a user.

A virtual computer must fulfill three requirements [3] to be considered as an equivalent to a physical computer:

1. *Fidelity or equivalence.* Any program that is executed within a virtual compute environment produces the same result as it would if executed on a computer without the virtualization environment.
2. *Efficiency.* Most instructions in a program are executed without any intervention by the virtualization system (i.e., mainly by the hardware itself such that a high performance is achieved). This implies that native instructions of the CPU are also native instructions of the application being executed.
3. *Resource control.* Within the machine it must be not possible for any running application to affect the other running applications resources (e.g., changing their memory content or interfering with each other's I/O).

These requirements ensure the correct (fidelity; as expected and repeatable), safe (resource control; no one can influence the result), and responsive (efficiency; higher performance) execution of programs on a virtual computer.

## 3.2 Managing the Virtual Resources: Resource Control and Efficiency

The number and diversity of resources in virtual computers requires a robust mechanism to manage access to their virtual resources. The hypervisor or virtual machine monitor is a software system that controls the access to virtual resources and mapping of requests to their physical resources in virtual computers.

The term hypervisor is derived from the notion of a supervisor for kernel in operating systems. Hypervisors are categorized as type-1 and type-2. Type-1 hypervisors run on top of hardware whereas type-2 hypervisors might be executed as a process by an operating system (see Figure 3.2). There are two additional classes of virtualization: containers and emerging unikernel-based VMs. These will be discussed later. For now, we focus on the introduction of heavyweight virtualization, which implies running VMs.

The type-1 hypervisors replicate closely (i.e., without changing or increasing the functional capabilities) the physical hardware and resources. The hypervisors are efficient according to [3] since they are not offering new functions from the resource to the guest. The type-1 hypervisors have major focus on strong resource access control.

Type-2 hypervisors are run on top of an operating system and interact with the host operating system resources management. This introduces unavoidable latency since all the hypervisor's action and the work of every VM must pass through the host operating system (OS). However, these hypervisors can implement additional virtual resource functionality, which is not available in the physical resources or to the guest (e.g., different memory management or even additional CPU instructions). Today the most common in practice is type-2.

## 3.3 A Brief History of Virtualization Concepts

The term *virtualization* in computer systems is difficult to define since many specific technologies are considered. Hence, we believe that it is beneficial to provide an overview and understanding of how virtualization concepts and virtualization techniques in computing and networking have developed and how

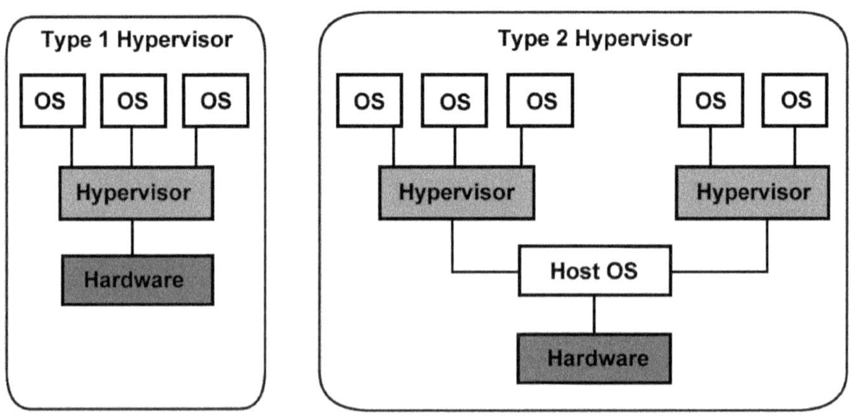


**Figure 3.2** Hypervisor types.

they relate. Furthermore, this should make it evident how the different aspects of virtualization in network and RAN engineering are addressed in this book.

The development of virtualization technologies for computing hardware and communication systems can be separated into three different streams of development with different activity levels. The streams are depicted in Figure 3.3. They overlap and influence each other.

*Stream 1:* Computer and OS virtualization and advanced operating systems with the aim of simplifying the sharing of computing resources.

*Stream 2:* Cloud computing for efficiently bundling and offering huge amounts of fine-grained computing resources and which are accessible from the outside through public networks for multi-tenant application usage. (Multitenancy is two or more unique users/customers sharing elements of a single system or server; think apartment dwellings.)

*Stream 3:* Network virtualization for adapting communication network and computing technologies to current and future CSP needs.

## 3.4 Virtualization Through the Ages

### 3.4.1 The Early Years: Computer and OS Virtualization

#### 1960s: Usefulness and Utilization

In 1961, researchers at the Massachusetts Institute of Technology (MIT) demonstrated for the first time the sharing of a computer using a Compatible Time-Sharing System (CTSS) operating system. This success laid the foundation for the Project on Mathematics and Computation (MAC) project, which aimed at making computing a commodity, like the access to water or electricity. The MAC results allowed the system to efficiently run multiple applications on a single mainframe computer.

Computer virtualization research in the second half of the 1960s focused on maturing the system's utilization. In 1964, the development of the Multiplexed Information and Computing Service (MULTICS) began. This was the first time-sharing operating system using single-level memory. One of the first deliveries of a commercial MULTICS-based computer, General Electric's 645 systems, occurred in 1967, and in 1968 IBM released the first version of its CP-67 control program as part of its Control Program/Cambridge Monitor System (CP/CMS) operating systems. In CP-67 one was able to create, monitor, and manage VMs on IBM's System/360 family of computers. By 1972, CP-67 was a robust, stable system running on 44 systems and could support up to 60 time-sharing users on an IBM S/360-67.

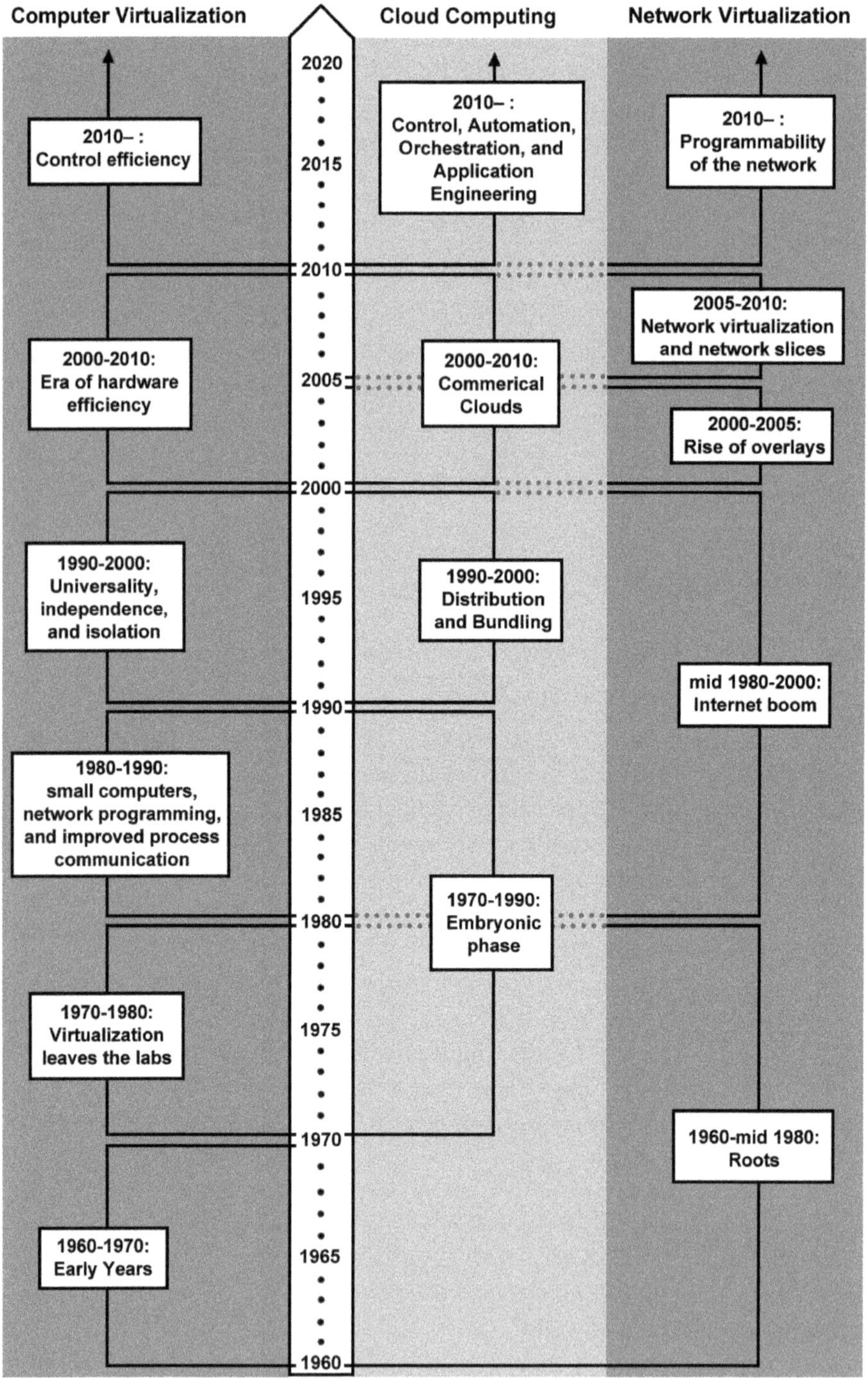


**Figure 3.3** Timeline of virtualization milestones.

### 3.4.2 The Second Decade of Virtualization: Virtualization Leaves the Research Labs

#### 1970s: Commercialization and Industrialization

The next period in computer virtualization is characterized by efforts to commercialize the systems. A fundamental course of action was IBM's decision in 1969 to unbundle computer software from computer hardware, the first known instance of what today we would call disaggregation. IBM adopted a marketing policy that charged separately for most systems' engineering activities, future computer programs, and customer education courses. This unbundling gave rise to a multibillion-dollar software and services industry.

Computer virtualization in the 1970s was characterized by its industrialization and major advances in programming concepts and operating systems. Major achievements were the development of advanced multitasking and multiuser operating systems and the related programming concepts such as modularization of programs [4], true procedural programming languages [5], processes and concurrency [6], and the development of service layers and layering concepts [7].

Major landmarks were the release of Version 2 Unix in 1972 and in 1978 the publication of a concise specification of the C programming language by Brian Kernighan and Dennis Ritchie [8]. These ideas were developed at AT&T Bell Labs and were the result of work to increase the utility of the newly developed PDP7 and PDP11 minicomputers by Digital Equipment Corporation (DEC). The development of the Unix Operating System was initiated to address shortcoming in the MULTICS system. While MULTICS enabled structured programming, it lacked features for easy management of parallel processes and the ability to develop programs concurrently, specifically the support for multiple usable shells (command line interpreters) or coordinated access for multiple and parallel users to stored files. Unix provided the Unix shell, which is command line interface (CLI) that allows interactive control of the execution of the system as well as automating this control using shell scripts. In addition, Unix implemented a hierarchical file system that related files and users and defined functional abstractions describing the user's abilities to create, read/retrieve, update, and delete/destroy files (CRUD).

In parallel, IBM developed the market for operating systems and software for their larger mainframe systems and, in turn, improved the industrialization of computers (i.e., their ability to generate business in areas other than their own functionality, especially in manufacturing). In August 1972 IBM announced the VM/370 operating system for its successful IBM System/370 hardware. VM/370 was a new implementation of CP/CMS.

### 3.4.3 Smaller Computers Join the Fray

In 1974, an independent software developer released a control program/monitor (CP/M) as a mass-market OS for Intel 8080/85-based microcomputers. This event can be considered as a reaction to the increased industrialization involving very large hardware. It is a focus on small, affordable, less risky, and more widespread computer systems.

Other important developments in the 1970s that intertwined with computer virtualization is the advancement of programmability concepts such as compilers, interprocess communication (including socket), and layering and service models. Using compilers, developers can run just about any program on any platform. Compilers translate code written in one language (the source language, normally human-readable) into another language (the target language, normally a machine language that while human-readable, is not easily comprehended by many humans). These high-level tools and compilers enabled the decoupling of hardware and software in the development process. Moreover, they permit the transfer and adaptation of code to a specific platform. However, it was still required that users compile all the software on the platform they wished to run on.

### 3.4.4 Processes Start Talking to Each Other

Interprocess communication (IPC) is a mechanism in an operating system to allow the processes to exchange shared data between two or more running processes. IPC defines a syntax (command and packet formats) and semantics (meaning and behavior) of data exchange. An important IPC mechanism is the concept of a network socket. A socket is a software structure within a network node of a computer network that serves as an endpoint for sending and receiving data across the network. The Berkeley sockets were released in 1983 as part of 4.2BSD Unix and represent an abstract representation (handle) for the local endpoint of a network communication path. The modularization of programs and use of procedural programming was initially outlined by Dijkstra's seminal work, previously in [6]. The split of programs into separate editable modules and procedures, together with the use of IPC, led to the understanding and development of layering and service models for computers and networks. Like procedures, services enable in *distributed systems* access to one or more capabilities where the admittance is provided using a prescribed interface and is exercised consistent with constraints and policies specified by a descriptor, as in the OASIS Reference Model for Service Oriented Architecture 1.0 (www.oasis-open.org). The service concept enables the transition from monolithic, aggregated, and centralized computer architectures to decentralized, heterogenous, open, and distributed computing systems. The Open Systems Interconnection model (OSI model), previously cited in [5], is a concept that characterizes and largely standardizes the communication functions of a telecommunication system without the need to consider the underlying internal structure and technology. The goal of the model is to permit and increase interoperability of diverse communication systems with standard communication protocols. One might be wondering why this is even required. In the early days of the computer era, different machines, often from different manufacturers, would have different physical characteristics that could render the ability to exchange information accurately and reliably impossible. While these machines all used binary (ones and zeros) the native structure of information stored in memory was often the first difference. The systems fundamentally were constructed with memory building blocks of eight bits, or octets. One can only represent 256 values in this amount of memory, not a very useful volume of data. Building a larger data structure out of two octets or 16 bits results in 65,536 possible values. While a data protocol exchange between two machines might be octet-based, if one machine is constructed using a 16-bit value, it could place the low-order octet and the low-order bits first in the address space (i.e., odd memory locations) and a second machine might have been designed with the high-order octet second in the address space (i.e., odd memory locations). One might think of the decision of making cars right-hand or left-hand drive; neither is a truly correct decision, it's just a choice and both choices have been made in different countries. This led to machines being called big ended or small ended. Now back to the point, if a big-ended machine was to send a 16-bit value over a connection to a small-ended machine, how would the small-ended machine know the originator was big ended? Trick question, it doesn't, so protocols were developed to accommodate this exchange. The ISO model allows the exchange through a network (the network is only aware of and managing the transfer up through the network layer) consisting of multiple nodes and to account for differences in machine designs and recover from errors (when desired).

### 3.4.5 Democratizing Computing in the 1980s

While virtualization was eventually not in the focus of computer and operating system research, the decade of the 1980s was characterized by the democratization of computing. The major event in this phase was the 1981 introduction of the IBM PC, which used a cost-efficient and open-hardware architecture and a separately licensed operating system—enter Intel and Microsoft. Four decades later the ubiquitous x86 computer architecture is not only in personal computers, laptops, and game consoles but also dominates the workstation and cloud server markets [9]. The popularity and new affordability of computing hardware reduced the need for virtualization techniques.

### 3.4.6 1990s: Universality and Independence

The 1990s reveal a renewal in the interest of virtualizing computers. Researchers at Sun Micro Systems were, in the first half of the 1990, looking for a better way to write and run applications than the established methods of the day. These processes were still very hardware-dependent and OS-oriented as well as focused on the C/C++ language and its application programming interfaces (APIs). In 1995, Sun introduced the world to a new language, Java (www.java.com). Java allows applications to be written once and later executed on any computer that has the Java Run-time Environment (JRE) installed. The JRE is a free application that uses just-in-time compilation (JIT) and provides the Java Virtual Machine (JVM), which is responsible for the execution of the Java byte code. In addition, the Java Development Kit (JDK) became available in 1996, which encompasses the JRE, a Java compiler, and additional developer tools such as a debugger and documentation generator.

JVM enabled the execution of Java code independent of the hardware architecture, bringing an unprecedented level of computer diversity. Java was not perfect, however. The performance and the resource efficiency of JVM was limited and due to the design of the memory management process, something called "garbage collection" resulted in programs that did not have predictable runtime or stable latency. In addition, JIT compilation leads to delays in start-time and the complex translation from byte code into assembler code is required each time a program is started.

### 3.4.7 2000: The Era of Hardware Efficiency

In 1999 VMware introduced its VMware Workstation (www.vmware.com), in which a central component was a hosted hypervisor. The VM in VMware stands for virtual machine, and the VMWare hypervisor was a type-2 hypervisor capable of running on top of Linux and Windows operating systems on x86 machines. This was the first commercial system that allowed users to define VMs on a single physical machine and use them simultaneously. Each VM can execute efficiently its own operating system such as Microsoft Windows, Linux, or BSD.

In 2003, a team of researchers from the University of Cambridge published Xen [10], a type-1 hypervisor allowing multiple VMs with their own OS to be executed on the same computer hardware concurrently. Xen is a free and open-source software project subject to the requirements of the GNU General Public License (GPL). It is included in virtually all major Linux distributions such as Debian, SUSE, or Ubuntu. Xen offers five approaches to run guest operating systems: hardware virtual machine (HVM), HVM with PV drivers (PVHVM, paravirtualization with full hardware virtualization), PV in an HVM container (PVH), and paravirtualization (PV).

To increase the usability and efficiency of OS virtualization the major CPU manufacturers introduced dedicated hardware support for virtualization in their CPUs. In 2005, Intel introduced the VT-x technology for its widespread x86 processors and in 2006 AMD announces AMD-V as its version of hardware support for virtualization.

The efficiency gained by improved hardware support was highly welcomed in the 2000s and sparked additional work to increase the efficiency of the virtualization concepts. This led to the development of the container concept (i.e., the specification of a method for running multiple isolated Linux systems (containers) on a control host using a single Linux kernel). The first containerization system was Linux containers (LXC). It provided functions to efficiently limit and prioritize the access to resources (CPU, memory, I/O, network, etc.) and isolation of the namespace that permits complete isolation of an application's view of the operating system, including process trees, networking addresses, user IDs, and mounted files systems. The container concept became more usable with the introduction of the Docker system in 2013. Docker enables OS virtualization delivered by software packages (denoted Docker containers).

The vision of the original VNF paper envisioned the interesting workloads running in virtual machines, and since that time containers have gained favor; in the near future, processes within containers will become more common. Ultimately several years in the future unikernel-based workloads are likely to become the standard for implementation. Figure 3.4 depicts this progression.

### 3.4.8 2010: Control Efficiency

The latest breakthrough advance in virtualization is the microkernel (often abbreviated as $\mu$-kernel). A microkernel is the near-minimum amount of software

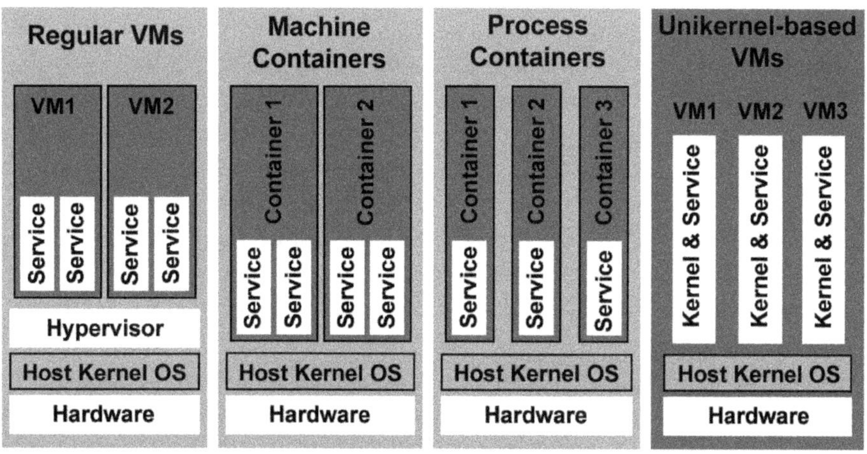


**Figure 3.4** Host virtualization options: past, present, and future.

that is needed to implement an operating system. These mechanisms include low-level address space management, thread management, and IPC. The support of microkernels might enable even smaller virtual representations of computers and might also enable more widespread use of virtual and specialist computers (e.g., in Internet of Things (IoT) systems) or for splitting software into even smaller pieces that can be run concurrently at distributed locations.

## 3.5 Cloud Computing

Let's start at the beginning of what today is commonly known as cloud computing.

### 3.5.1 1970–1980: The Embryonic Phase

The metaphor of a cloud was used early in the original Advanced Research Projects Agency Network (ARPANET) to describe the compute capability of the network rather than to detail the actual connectivity among the end hosts. The metaphor makes it evident that the utility of the network was not only for users' communication services and the general exchanging of information, but also to enable remote access to computing capabilities.

The early and embryonic version of cloud computing in the ARPANET was mainly focused on mainframes as the major computing infrastructure element. Common were the General Electric GE-645, DEC PDP-7/PDP-11 and IBM System/370. The on-demand access to these infrastructures was permitted by the contemporary networks. The exchange of any large volume of data among these systems and users was, to be kind, limited. This was because of the restricted efficiency, performance, and throughput of the available communication networks in the 1970s and 1980s. The network connections of the day typically consisting of dedicated point-to-point circuits for data communication and interconnected by network nodes that had limited compute for handling packets. The limitation in exchanging data and processing packets also made it difficult to exchange information for the system management and control functions such as systemwide user identity or access management.

The second half of the 1980s witnessed an increase in availability of computing capabilities by the democratization of computing hardware and the improved understanding of ARPANET transmission and control capabilities. This rapid growth transformed this research network into today's internet. In 1984, researchers and engineers at Sun Microsystems coined the phrase "The Network is the Computer" [11]. The concept of this phrase was to have the network within the core of the computer. The vision at Sun at the time was an interconnected world based on open and shared standards for computing and communication.

### 3.5.2 1990: Distributed and Bundling

During the 1990s, the capabilities of data communication networks improved dramatically. Using the circuit switched model, virtual private networks (VPNs) became increasingly available and were offered as a service by CSPs. These enabled easy access control through service validation, and permitted in turn, a higher level of security for the compute infrastructure. This period also saw the rapid growth of the use of the IP in communication networks. IP and related technologies improved the speed, efficiency, and programmability of communication networks and began to reduce the cost of the associated network transport.

As a result, the 1990s reveal increasingly the use of the term "cloud" to refer to the platforms for distributed computing. These platforms were available in various forms such as grid computing, mainframe computer, or even client-server models; in general, any distributed application that separated service providers (servers) and service requestors (clients). In addition, it was more commonly understood that the elegance of distributed platforms is found in their capability to compute functions in different places. As outlined by Andy Hertzfeld in 1994, "...we now have the entire Cloud out there, a single program can go and travel to many different sources of information and create a sort of a virtual service" [12]. This statement is considered by many as the starting call for bundling a vast amount of computing resources in the cloud and making them accessible.

### 3.5.3 2000: The Cloud Becomes a Commercial Offering

A major milestone in practical cloud computing occurred in 2002 with the founding of Amazon Web Services (AWS) as a subsidiary of Amazon. A motivation for AWS was the need to scale engineering and operation for Amazon's e-commerce-as-a-service platform (e.g., such that users can build their own webstores). A further incentive was provided by the insight that Amazon's IT platform was spending 70% of its time idle and that the unused resources might be turned into user-facing compute services that generate revenue and income to Amazon when third-party developers use the infrastructure. AWS created a family of on-demand cloud computing platforms and APIs and provided them to individuals, companies, and governments on a metered and pay-as-you-go basis. In March 2006, AWS released its Simple Storage Service (S3) followed by the popular Elastic Compute Cloud (EC2) in August of the same year. Elastic

Cloud platforms became popular—Google announced its Google App Engine in 2008 and Microsoft launched its Azure Cloud platform in 2010.

The first decade of the 2000s was also characterized by the transition from compute, web, or file hosting infrastructure to a taxonomy of cloud computing services. The notion of "X-as-a-service" was adopted to categorize and characterize the capabilities of the services and the APIs offered to customers. In the meantime, a large number of specific X-as-a-service concepts have emerged. The most important ones are infrastructure-as-a-service (IaaS; high-level APIs used to control various low-level details of underlying physical computing and network infrastructure), platform-as-a-service (PaaS, or application platform as a service; management and programming of a modular bundle computing platform related to one or more applications and without the need to build and operate an infrastructure), and software-as-a-service (SaaS; the cloud platform acts like a delivery system and licensing model for an application that is centrally hosted, often called on-demand-software).

### 3.5.4 2010s: Control, Automation, Orchestration, and Application Engineering

The recent decade in cloud computing started in 2010 with availability of OpenStack (www.openstack.org). OpenStack is an open standard cloud computing platform that can be characterized as an IaaS system. OpenStack constitutes a control framework for cloud that enables the simple creation of public and private clouds where virtual servers and other resources are made available to users. It enables the operators of those clouds to efficiently manage them.

The widespread availability of control frameworks such as OpenStack triggered the next milestone in cloud computing. This step is the upswing in tools for automated, concise, and large-scale management and configuration (i.e., tools for the orchestration of cloud computing infrastructures). In 2014, Google released Kubernetes as a tool for automating computer application deployment, scaling, and management in their cloud platform. The tool takes advantage of the increased use of Docker. Kubernetes allows for the efficient programming of the deployment of Docker containers and the configuration of servers in data centers running virtualization by containerization. While Docker provides an open standard for packaging and distributing containerized applications, the potential complexities in coordination and configuration can add up fast. Kubernetes provides an API to control how and where those containers will be run. In addition, it provides APIs for automating the configuration of Kubernetes clusters. The term infrastructure-as-code (IAC) was coined during this decade. The term characterizes the process of managing computational resources in cloud platforms through machine-readable definition files instead of making manual configuration changes or using interactive configuration tools. The elimination of human interaction in configuration tasks increases the scalability, decreases the operational cost, and makes these infrastructures even more reliable since automated configuration actions are provable and faster. Subsequently, many IAC tools have been developed, including Puppet (2005), Chef (2009), Ansible (2012), and Terraform (2014).

Finally, the rise of automation and orchestration capabilities leads to another, and probably not final, revolution of cloud computing: the application of DevOps. As explained in Chapter 1, the term DevOps is a combination of "development" and "operations," and it describes a methodology in software engineering and is defined as a "a set of practices (in software design) intended to reduce the time between committing a change to a (software) system and the change being placed into normal production, while ensuring high quality" [13]. Cloud application design by DevOps is often characterized by three major principles: shared ownership of the software, workflow automation (e.g., through automated deployment and orchestration in the cloud), and rapid feedback to fix bugs and improve the cloud application. DevOps has its roots in the Telecommunications Information Networking Architecture Consortium (TINA-C). TINA-C defined a model of a service lifecycle that combined software development with (telecom) service operations [14]. The use of DevOps and the need for coordination of the development of open-source tools for virtualization and orchestration led to the founding of the Cloud Native Computing Foundation (CNCF) in 2015 (www.cncf.io).

## 3.6 Network Virtualization

The concept of network virtualization is a rather recently defined concept in the history timeline of virtualization and the cloud. However, its roots date back to the origins of today's internet and of the IP protocol stack as well as other developments in the area of programmable networks [15].

### 3.6.1 1960–Mid-1980: Roots and Programmability of Distributed Computing

The original four-node ARPANET from 1969 was conceived as a logical overlay network for packet relaying on top of an automated and connection-oriented telephone network, and here we start to see the tie into computing, virtualization, and networking. The network consisted of dedicated leased lines (leased from the telephone network providers) and were interconnected by interface message processors (IMPs), the ancestor to today's routers. The 1970s and 1980s were characterized by efforts to increase the efficiency of the packet forwarding and the simplification of the programmability of the data exchange between end hosts for distributed applications. This resulted in the well-known development of the TCP/IP stack, its related protocols, and the specification of the socket model for the data exchange mentioned earlier.

### 3.6.2 Mid-1980–2000: The Internet Boom

The need for abstraction was at first still low. Networking research focused mostly on the understanding of host-to-host communication and the roles of the host. The client/server model became the standard role model. An important understanding and interpretation of a logical virtual network took place when the World Wide Web, and its related technologies such as HyperText Markup Language (HTML) and hyperlinks, where introduced. The internet was now perceived as a virtual network of applications, services, and data with interconnections established by hyperlinks and described by Uniform Resource Locators (URLs).

The evolution of the Web took place in waves. The first wave was Web 1.0—the webservers provided content (producer mode) and the users accessed the content via web clients denoted as browsers (consumer mode).

The increased usefulness and economic importance of the internet and the Web became evident with the creation of Commercial Internet Exchange (CIX), which interconnected various providers' networks to increase the network-of-networks effect of the internet system.

### 3.6.3 2000–2005: Powerful Application Overlays and Ossification of the Internet

Web 2.0 blurred the boundary between content producers and content consumers. The nodes became "prosumers," a combination of "producer" and "consumer," due to the equalness of the nodes that were described as peers. This view gave rise in the early 2000s of powerful peer-to-peer (P2P) application overlays. These overlays were abstract and logical networks on top of physical ones. They were established as logical links between end hosts and the end hosts were acting as interconnecting nodes. A first major successful P2P overlay was Gnutella, which was released in March 2000. It is a P2P file-sharing overlay, and its purpose was to advertise, locate, and exchange mp3 files. In subsequent years, a significant number of P2P filesharing overlays and applications were released: eDonkey (2000), Kazaa (2001), and BitTorrent (2001).

It becomes evident that the capability of P2P technologies is not only limited to filesharing but could also be applied to control tasks in networks and telecommunication services. The most prominent example was the original version of Skype, which was released in 2003. This version of Skype used the user's end hosts as a self-organizing set of index servers for locating users and as proxies to relay calls to the current location or host of the user. In this way, Skype solved the task of locating roaming users as well as avoiding very large index servers (e.g., for mobility management). The reliability of Skype was based on the assumption that the index server can join or leave voluntarily and the protocols for self-organization among the index servers ensures the synchronization of location information. Additional academic research on using P2P technologies for network control is outlined and summarized in [16].

Academic research during this time revealed that new routing and resilience models applied in application overlays can outperform the ones applied in physical networks. A major study was Resilient Overlay Networks (RON) in 2002 [17], which defined an architecture that "...allows distributed Internet applications to detect and recover from path outages and periods of degraded performance within several seconds, improving over today's wide-area routing protocols that take at least several minutes to recover." Research on distributed hash tables (DHTs) showed that infrastructures that were using this data structure can efficiently perform highly complex services such as anycast, cooperative web caching, distributed file systems, domain name services, and content distribution systems. DHTs are superior in their scaling and can easily support millions of nodes due to their logarithmic runtime (e.g., when looking up nodes or key-value pairs). Notable DTHs of this era are Chord (2001), Pastry (2001), and Kademlia (2002).

### 3.6.4 2005–2010: Network Virtualization and Network Slices

In parallel, it became evident, due to its huge and rapidly expanding footprint, that it is next to impossible to introduce major new innovations and technologies in the contemporary internet, for example new routing or resilience mechanisms. It was concluded by a group of leading network researchers that were supported the National Science Foundation (NSF) that the *ossification* of the internet is a natural evolutionary stage [18]. This characteristic means that the internet's success created a molded and rigid status that

> ...in turn, created inertia that inhibits change. However, the problem is more acute in the context of network technologies because network technologies are shielded from effective competition by the deployment obstacles raised by the excessive cost of infrastructure and the need for agreement among a diverse collection of organizations with often competing interests. ....https://www.arl.wustl.edu/netv/main.html, 2005

Hence, network virtualization was suggested as a strategy for addressing the internet's standstill in major technologies. In such a virtualized network system, multiple virtual networks can coexist and are embedded on top of a shared substrate, again seen in [18]. The authors of this 2005 study observed that different virtual networks could provide alternative end-to-end packet delivery systems and could use different protocols and packet formats. Virtual networks, often denoted as slices, are implemented by virtual routers connected by virtual links. A major milestone in mastering network virtualization became the availability of the PlanetLab system [19]. PlanetLab was a global research network that supported the creation of new network services. It was set up as a testbed for computer networking and distributed systems research. Originally established in 2002, it grew to over 1,000 nodes at 500 sites worldwide at its height. While the nodes were physically managed locally, each experimenter's project were assigned a "slice" or virtual machine access to a subset of the nodes. The user management of the nodes was centralized, and user accounts were synchronized with a slice (i.e., the subset of nodes). PlanetLab resulted in the Global Environment for Network Innovations (GENI) project, which is a global but mainly North American distributed high-performance networks lab. It bundled networking research projects in its series of GENI Engineering Conferences (GECs).

While PlanetLab was a successful tool for networking and distributed systems research, it missed the advances in transmission technologies. The packet forwarding in virtual machines can apply and demonstrate complex algorithms, but the speed of the packet processing was limited by the available computing resources. As a result, networking researchers were searching for ways to take advantage of the new and fast packet forwarding technologies available in switches. The new switches provided very high packet throughput and the forwarding tables of those switches were configurable (originally by the internal switch protocols, such as spanning tree). This led to the development of OpenFlow (OF) [20]. OpenFlow is a communications protocol that gives access to the forwarding plane of a network switch or router over the network. It was initially intended to control Ethernet switches, updating internal flow-tables using a standardized interface to add and remove flow entries. OF uses a manager/agent model for management, where the OF controller acts as a manager and communicates with the switch as an agent and by using the OF protocol. The controller configures and controls the forwarding table and behavior of the switch. The manager/agent model also enables the separation of control plane (OF controller or set of OF controllers) and data plane (single or set of switches).

Finally, the term software-defined networking was coined at the end of the decade (2009), describing the programmability of the data plane [21].

### 3.6.5 2010: Programmability of the Network

Recent years have been characterized by increases in the understanding of the programmability of the data plane and the control plane. Data plane programmability advanced in 2013 with the definition of P4 [22]. P4 is a domain-specific programming language for controlling packet forwarding planes in networking devices. The P4 name stands for "Programming Protocol-Independent

Packet Processors" [22] where the language was specified. Specific programable silicon is usually required to support P4 control on the switch or network device.

The need for control plane programmability and for virtualization of network functions was outlined in the 2012 ETSI white paper "Network Functions Virtualization" (NFV) [23]. As discussed in Chapter 1, this paper describes the benefits, enablers, and challenges for the virtualization of network functions, as distinct from cloud and SDN, and if one has not yet read the relatively short text it is highly recommended at this time. The paper also discusses a rationale for encouraging collaboration, for both software and hardware development, to accelerate development and deployment of interoperable network control and network service solutions based on high-volume industry standard servers. This encouragement is to the entire ecosystem from silicon designers to network equipment manufacturers, independent software vendors, system integrators, and to the CSPs themselves.

Finally, the cloud-native network paradigm [24] was introduced at the end of the decade. In cloud-native networking, the (virtual) network functions or network services were implemented by the design principles of cloud-native applications (i.e., by interconnected software modules in the cloud, eventually denoted as a service chain).

## 3.7 Basic Objects and Data Structures for Network Virtualization

The discussion of virtualization techniques makes it evident that the technologies for the virtualization of networks are specific, complex, and diverse. Virtualization also aims at abstraction and generality. A solution to these contradicting aims is the view that a network is an object that achieves a function, denoted as the network function at large. The virtualization of a network aims at achieving this function by its abstraction and the support for computing it. These network functions can be considered as objects that eventually can be replaced or overloaded by other functionalities or objects, like function overloading, a common concept in object-oriented programming (OOP).

The basic data structures for network virtualization can be derived initially from the original specification of virtual networks and used to overcome the internet ossification impasse. The taxonomy of the data structures comprises four categories in defining basic network characteristics: network graph (topology), the names used to identify nodes in the network (addresses), the mechanism to find nodes in a graph (routing), and the resources and mechanisms for data forwarding (resource management).

### 3.7.1 Network Topology

The network topology is described by the information and data contained in a graph. A graph is typically a representation in a two-dimensional plane, where the vertices of the graph are represented by distinct points and the edges by arcs joining the corresponding pairs of points. Network virtualization considers the virtual topology of a network as the data object that describe the network that is separated from the hardware.

### 3.7.2 Addressing

A network address is a data structure and the identifier for a node or host on a telecommunications network. The most prominent identifiers are IP addresses. A network address serves two main functions: host or network interface identification and location addressing.

More elaborate addressing can be applied beyond what is possible with the IP addressing. One technique being used today is named data networking (NDN) paradigm. NDNs are opaque to the network, which allows each application to choose the naming scheme that fits its needs, and naming can thus evolve independently from the network [25].

DHTs use an abstract keyspace, such as the set of 160-bit strings. A keyspace partitioning scheme for a specific DHT algorithm splits the ownership of this keyspace among the participating nodes. An overlay network then connects the nodes, allowing them to find the owner of any given key in the keyspace [26].

The network addresses in a virtual network might be defined independently from other virtual networks that are supported in the same physical structure. Similarly, the addresses in a virtual network might be separated from the physical addresses. Indirection and naming services like the Internet Indirection Infrastructure (i3) [27] or Domain Name System (DNS) can perform the translation between physical and virtual addresses.

### 3.7.3 Routing

Routing is the process of selecting a path for traffic in a network or between or across multiple networks. The virtualizable objects for the routing are the routing function (i.e., the routing algorithm itself) and the routing tables as data structures. Hereby, the routing algorithm is part of the control plane, and the routing table is part of the data plane.

### 3.7.4 Resource Management

Finally, the resource management for a single link interconnecting two end points constitutes the last category of data structures. This management can comprise packet scheduling on a (virtual) link, packet forwarding or any flow control, and media access mechanisms. The virtualizable functions are the algorithms for the resource management, and the virtualizable data structures are resources descriptors and the status information for the link.

## 3.8 Summary

Network virtualization concepts have been developed over many years and have recently become practicable and able to accommodate new technologies, including the large number of technologies emerging from cloud computing. However, usage and business scenarios will probably be applied more realistically on a longer timescale. However, the ongoing rapid development requires a detailed understanding of current virtualization technologies as well as the capability to engineer networks such that they embed these techniques into their core functions. One also needs to be able judge the performance and economic viability of a virtual network design, topics that will be discussed in subsequent chapters.

## References

[1] https://www.collinsdictionary.com/dictionary/english/virtual.

[2] Faynberg, I., H. Lu, and D. Skuler, *Cloud Computing: Business Trends and Technologies,* Chichester, UK: Wiley-Blackwell, 2016.

[3] Popek, G. J., and R. P. Goldberg, "Formal Requirements for Virtualizable Third Generation Architectures," *Commun. ACM,* Vol. 17, No. 7, July 1974, pp. 412–421, doi: 10.1145/361011.361073.

[4] Barnett, T. O., and L. L. Constantine (eds.), *Modular Programming: Proceedings of a National Symposium,* Information & Systems Institute, 1968.

[5] Zimmermann, H., "OSI Reference Model-The ISO model of Architecture for Open Systems Interconnection," *IEEE Transactions on Communications,* Vol. 28, No. 4, 1980, pp. 425–432.

[6] Dijkstra, E. W., "Letters to the Editor: Go To Statement Considered Harmful," *Communications of the ACM,* Vol. 11, No. 3, 1968, pp. 147–148.

[7] Dijkstra, E.W., "Cooperating Sequential Processes," in *Programming Languages,* F. Genuys (ed.), Academic Press, 1968.

[8] Kernigan, B. W., and D. M. Richie, *The C Programming Language,* Upper Saddle River, NJ: Prentice-Hall, 1988.

[9] Brandon, J., "The Cloud Beyond x86: How Old Architectures are Making a Comeback," April 15, 2015, https://icloud.pe/blog/the-cloud-beyond-x86-how-old-architectures-are-making-a-comeback.

[10] Barham, P., et al. "Xen and the Art of Virtualization," *ACM SIGOPS Operating Systems Review*, Vol. 37, No. 5, 2003, pp. 164–177.

[11] Perry, T. S., "Does the Repurposing of Sun Microsystems' Slogan Honor History, or Step on It?" *IEEE Spectrum*, July 30, 2019.

[12] Levy, S., "Bill and Andy's Excellent Adventure II," *Wired*, April 1994, p. 107.

[13] Bass, L., I. Weber, and L. Zhu, *DevOps: A Software Architect's Perspective*, New York: Addison-Wesley Professional, 2015.

[14] Chapman, M., and N. Gatti, "A Model of a Service Life Cycle," *Proceedings of TINA '93*, September 1993, pp. I-205–I-215.

[15] Feamster, N., J. Rexford, and E. Zegura, "The Road to SDN: an Intellectual History of Programmable Networks," *ACM SIGCOMM Computer Communication Review*, Vol. 44, No. 2, 2014, pp. 87–98.

[16] Tutschku, K., *Peer-to-Peer Service Overlays–Capitalizing on P2P Technology for the Design of the Future Internet*, Habilitation Thesis, University of Würzburg, Institute for Computer Science, Department of Distributed Systems (Informatik III), June 2008.

[17] Andersen, D., et al., "Resilient Overlay Networks," *ACM SIGCOMM Computer Communication Review*, Vol. 32, No. 1, 2002, p. 66.

[18] Anderson, T., L. Peterson, S. Shenker, and J. Turner, "Overcoming Barriers to Disruptive Innovation in Networking," *Report of National Science Foundation Workshop*, January 2005.

[19] PlanetLab, https://planetlab.cs.princeton.edu/about.html.

[20] Open Networking Foundation, https://opennetworking.org.

[21] Software-Defined Networking (SDN) definition: https://opennetworking.org/sdn-definition.

[22] P4 Open Source Programming Language, https://p4.org.

[23] ETSI Network Function Virtualization, https://portal.etsi.org/NFV/NFV_White_Paper2.pdf.

[24] Persson, H., and H. Kassaei, "Cloud-Native Application Design–In the Telecom Domain," *Ericsson Technology Review*, June 2019.

[25] Named Data Networking, https://named-data.net/project.

[26] Wiley, B., "Distributed Hash Tables–Part 1," *Linux Journal*, October 2003.

[27] Stoica, I., D. Adkins, S. Zhuang, S. Shenker, and S. Surana, "Internet Indirection Infrastructure," in *Proc. of the ACM Sigcomm 2002 Conference*, Pittsburgh, PA, August 2002.

# 4

# Data Plane Virtualization and Programmability for Mobile Networks

## 4.1 Data Plane Acceleration with OpenFlow and P4

This chapter will discuss and detail the concepts and current two main technologies for data plane virtualization in 5G and B5G mobile communication systems. The aim is to cover programmable packet forwarding in the data plane of networks to facilitate virtual communications networks, specifically those found in the modern CSP mobile network. The chapter will focus on the major programmability concepts and protocols that have been developed in last decade: OpenFlow and P4. Before explaining these protocols, we will first outline the concept of network functions and the separation between control and data plane in mobile networks.

### 4.1.1 Context for Acceleration

4G, 5G, and future 6G networks will be woven into the daily lives of nearly every human, business, and industrial application. Mobile communication networks have a much higher degree of complexity than the original internet architecture [1] and comprise many more network functions (NFs) than foreseen in the original internet.

The characteristics realized in the mobile network are built around NFs. These NFs evolved as the network evolved and at first supported the need to support roaming users and devices, as well as manage the use of RF transmissions as the main mode to connect end devices to the network. In addition, in the mobile network there is the requirement to bill customers for the use of the system, measuring minutes and use or packets transported (in both directions). The original internet concept didn't focus on mechanisms to support mobility or accounting and billing. There is another key difference and that is the requirement for legal compliance (sometimes called lawful intercept) and for the CSP to retain transaction records.

As a result, the modern and public mobile network is designed around the required functionality for mobility and commercialization. While the internet focuses on forwarding of data and the interconnection of networks, mobile networks require a more structured architecture with additional functionality. A common approach is to consider *functions* as a universal structuring scheme. The concept of function is not completely new in the CSP; it is very similar to the concepts of procedural [2] and functional programming [3] found in computer science. One can distinguish in networks and in general in mobile networks both *network and service control functions* (i.e., functions that manage and coordinate the network, data forwarding, and functions providing the services offered by the network) and *data plane functions* that perform the data forwarding. This separation in the CSP environment is often referred to as CUPS, as previously discussed in Chapter 2.

A simplified 5G system architecture of network control functions and data plane functions is shown in Figure 4.1. In reality, the complete architecture is significantly more complex. The interconnects of the functional modules depend on switches, routers, and other traffic transport and managing devices. The scale and redundancy of the servers in a production network are also often not shown in these high-level figures. To understand the full set of network

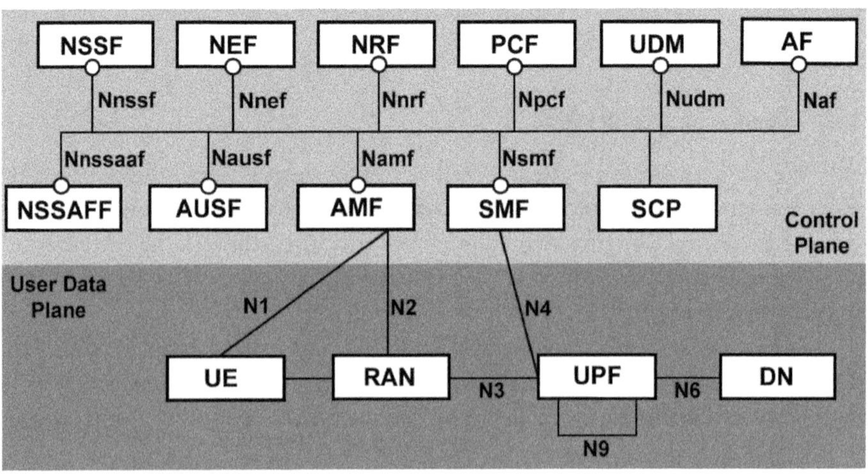


**Figure 4.1** 5G system architecture separation between data plane and control plane.

functions in 4G/5G systems, the reader is referred to more specific literature on 4G and 5G mobile networks [4] and to the 3GPP standards themselves [5].

Figure 4.1 shows the two major parts of a 5G mobile communication network: (a) the data plane and (b) the control plane. The data plane functionality in a 5G system between the user equipment and the mobile core network, using the UPF. The control plane of a 5G system might comprise *network control plane services* such as network slice selection function (NSSF), network exposure function (NEF), network repository function (NRF), policy control function (PCF), unified data management (UDM), application function (AF), AMF, authentication server function (AUSF), network slice-specific authentication and authorization function (NSSAAF), SMF, and service communication proxy (SCP).

The separation of the data plane and the control plane is a design feature of the 5G standards and takes into consideration the functionality required in a mobile network to support devices that change locations (e.g., roaming, the path of the connection may change during the life of any session). A second motivation for CUPS is to address increased traffic volume, higher reliability requirements, and to provide predictability in networks. The demand for very fast control in certain network management functions has implications in traffic engineering (i.e., for the mechanisms that provide control over the paths used to deliver traffic through the network to the UE). The conventional internet architecture, from the earlier reference [1], although highly successful in enabling the distribution of very high data volumes, embodies very tight integration of the control and data in the routers and switches. Since the UE can be mobile, this mobility implies there may be different routes when there are active connections. IP mobility support for IPv4/IPv6, known as MobileIP, is found in the IETF RFCs 5944 and RFC6275. MobileIP supports seamless TCP connections between a moving device and a static one using mechanisms in the IP layer. MobileIP, however, couldn't demonstrate in practice the support for very high data rates. MobileIP is also challenged supporting very low packet delay or delay variation, also known as jitter. The MobileIP protocols and its mobility management functions require a large amount of computation for the data packets and this is done in the upper protocol layers; consequently, acceleration of the packets may be required.

Finally, the topic of distributed state management for data forwarding and routing in network requires a brief discussion. Local forwarding rules (i.e., the ones for relaying a packet from an ingress port to an egress port) can be efficiently interpreted and fulfilled by L2 switches with dedicated hardware capabilities. These switches can achieve very high throughput and very short packet delays. Logically centralized route controllers need to maintain replicas of the routing and network state information at one or more locations, called replicas. Reliability and resilience are thus enabled with these replicas at the risk of temporarily inconsistent state information. Internet routing has, however, demonstrated that replicas can eventually converge after learning the routing and topology information.

Specifying an optimal architecture for control (centralized vs decentralized) for a certain network and its application remains a challenge. An optimal architecture can depend on the applications and their control cycle times. It is becoming increasingly evident that a separation of the control plane from the data plane has positive implications for the control cycle time. In addition, this separation enables a flexibility to upgrade or to implement a specific control architecture. The rapid adoption of the separation of the data and control plane is evidenced by use of this concept in the standards for 4G and 5G mobile architectures being developed by ETSI and 3GPP [6, 7].

With this background established, it's time to discuss the first of the two current important protocols: OpenFlow.

## 4.2 OpenFlow

OpenFlow [8] emerged in 2008 from research. Existing control frameworks of the time were not aligned with mechanisms implementing control activities in the data plane.

OpenFlow focuses on the fast forwarding of packets (acceleration) within *flows* (flows are matched to fields found in the header of the IP frame and apply equally to TCP and UDP packets) and the rapid reconfiguration of the forwarding elements (i.e., of the switch) for handling these flows. OpenFlow provides a clear separation between data and control planes. The controller may also forward packets if necessary. An OpenFlow controller might have more functionality than a pure management element found in other protocols.

OpenFlow enables a balance between the vision of a fully programmable network and the required pragmatism of real-world implementations [9]. It is considered as the first applicable mechanism that aims at achieving simplified and elegant programmable high-speed networking.

OpenFlow's major architectural progress is to offload the forwarding elements from complex and potentially networkwide control functions and from the related control software. OpenFlow uses secure control channels between controllers and switches to improve security within the network.

The OpenFlow specification is maintained by the Open Networking Foundation (ONF; www.opennetworking.org). The subsequent description of OpenFlow is based on ONF's specification Version 1.5.1 (Wire Protocol 0x05) [9], the current generally available version. (Note that ONF maintains a Version 1.6 that is only available to ONF members.)

### 4.2.1 Flows

OpenFlow's simplicity arises from the consideration of packet flows and their fast and iterative handling. In OpenFlow, a flow is a sequence of packets identified between a source and a destination. Source and destination are not only specified by typical address concepts (i.e., IP addresses or MAC addresses) but also header fields within a packet, such as application port numbers. A flow is a much finer grained stream of packets than a path in typical internet routing, and in turn, is a more general abstraction.

OpenFlow allows for iterative handling of packets of a flow within a switch. The forwarding element may execute multiple consecutive actions on a packet (e.g., modifying headers and forwarding). This permits complex actions and modifications in addition to the forwarding of packets. An OpenFlow switch may implement parallel packet handling for packet handling in a flow, which may result in throughput increases.

### 4.2.2 Configuration

OpenFlow assumes that the patterns for matching flows and the associated forwarding actions can easily be (re-)configured during a switch's operation. A switch should not need to be rebooted, nor should a change in the flow table cause a longer interruption in the packet forwarding or the exchange of state information with other switches or forwarding elements. As a result, a switch does not require comprehensive state changes or networkwide update.

### 4.2.3 System Model and Pipeline

An OpenFlow system comprises at least one *OpenFlow switch and one or more controllers,* as shown in Figure 4.2. Packets received on ingress ports are extracted and processed by a sequence of one or more *flow tables within the switch.* The sequence is denoted as the *pipeline.* The logic in the tables performs lookup, matching, and forwarding logic. A table may also forward a packet to the controller for special handling. An OpenFlow switch also implements two additional tables, Group and Meter, to provide more specific processing and to monitor and control rates, respectively. OpenFlow also supports one or more Channel modules, which maintain the secure communication to the controllers.

An OpenFlow controller can add, update, and delete flow entries in the flow tables in a switch. The change may be *reactively* (i.e., in response to packets in a flow that was received from the switch) or the controller may do this *proactively* (i.e., by anticipating flows and a priori assumptions to handle the flow).

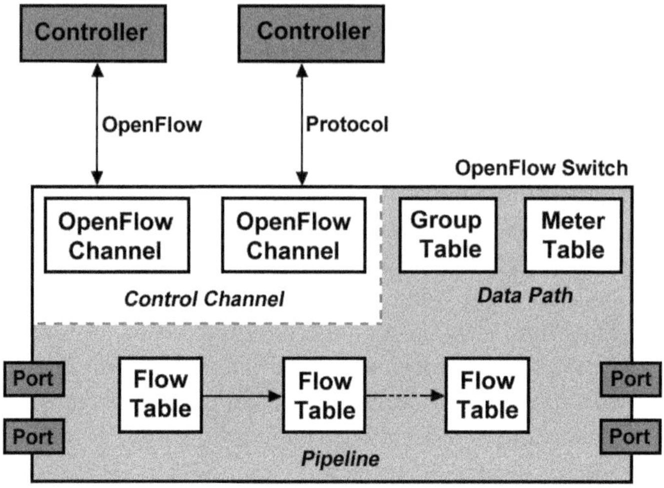


**Figure 4.2** OpenFlow's system model.

These two modes are often combined in real-world OpenFlow scenarios. If a controller received a data packet, it may return it back to the switch for further processing and forwarding, it may discard the packet, or the controller may provide the forwarding itself.

The matching of packets starts at the first table in a *pipeline* and continues through subsequent tables. A packet match with the *highest priority table entry* activates the further packet processing. Packets may be forwarded to the next table, toward an egress port or controller for additional treatment or dropped. Any forwarding may include additional actions on the packet.

### 4.2.4 Ports

Ports are the representations for the physical links of a switch. In addition, *logical ports* may facilitate switch-focused forwarding treatment (e.g., link aggregation, tunnels, or loopback interfaces).

Another important category of logical ports is the *reserved ports.* Reserved ports enable the interconnection of elements in the forwarding abstraction model (see below). Major and required reserved ports are CONTROLLER, TABLE, and ALL. The CONTROLLER port provides the control channel to the OpenFlow controllers. It is used to send and receive packets to and from the controller (i.e., for special handling by the controller). The TABLE port represents the start of a pipeline, and the ALL port is a placeholder for all physical ports. Other mandatory reserved ports are IN PORT; for example, any ingress port and ANY that is used when no port is specified (i.e., the port of last resort).

### 4.2.5 Group, Meters, and Counters

Flow actions may direct packets to the Group data structure. This structure contains ordered lists of actions on a packet that are executed when the packet is exiting the processing pipeline.

A meter will measure and control packet rates. It triggers a *meter band* exception if the packet rate or byte rate passing through the meter exceeds a predefined threshold. Counters are elements that accumulate statistics at various specific points of a pipeline (e.g., on a port or at a flow entry). Counters count the number of packets and bytes passing through an element. Other counter types may also be defined.

### 4.2.6 Forwarding Abstraction

A core feature of OpenFlow is the availability of fast-forwarding mechanisms, generalizing them into a model for a *forwarding abstraction.* The abstraction describes the generalized concept of how packets are passed through a switch. It also comprises the notions and general behavior of the steps doing the packet processing. OpenFlow's forwarding abstraction is depicted in Figure 4.3. It is built around the structure of a flow processing *pipeline* and its sequence of elements (e.g., the *Tables*). The tables have numbered identifiers, and the packets move through them from lower to higher IDs. (The table IDs create a rank ordered list that controls the pipeline flow.)

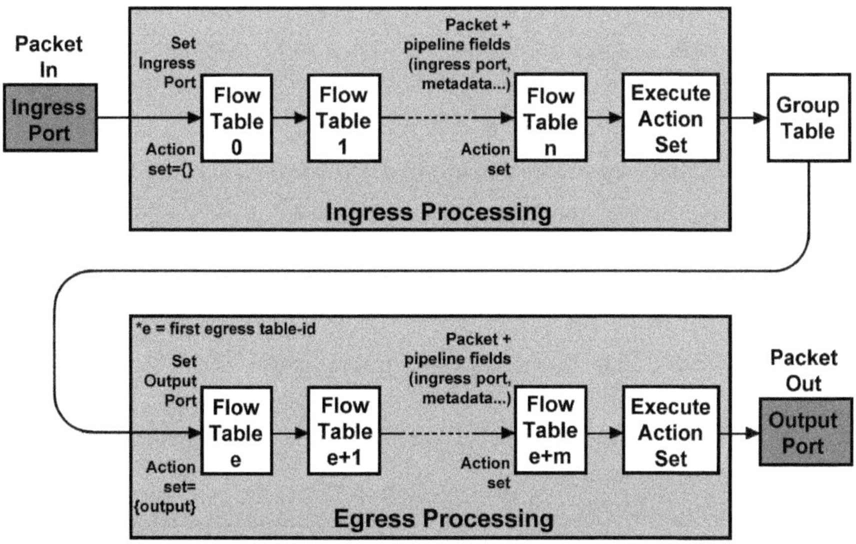


**Figure 4.3** Forwarding abstraction in OpenFlow.

OpenFlow distinguishes between *Ingress* and *Egress processing* pipelines. A pipeline is ended by the execution of an *Action Set* for a packet. An OpenFlow switch must support at least one Ingress pipeline, while an Egress pipeline is optional. The Action Set accompanies a packet during the journey through the pipeline. A *Table* consists of multiple rows and each row consists of pattern-matching logic and a set of instructions and actions to be executed upon detecting a match. When a packet arrives at a pipeline, headers are extracted and compared with the specified patterns for the first Table in the pipeline. A match in the table triggers the execution of instructions, including the update of the *Action Set* and any *metadata;* this information can be passed from one table to another. A key table entry instruction is to forward the packet to one of the next tables in the pipeline.

The OpenFlow specification requires compliant switches to support a minimal pipeline length of one. Such a minimal length might simplify the switch's implementation but also might limit its capabilities for flow processing (e.g., state-based changes and multiple changes on a packet are not possible). These switches may fail to be consumable in a real network.

Figure 4.4 shows a simplified ingress and egress flowchart through a table entry in a pipeline. The egress logic omits the Group data structure in its pipeline. A packet from an ingress port arrives at the pipeline, it is matched or handed over to the next table or cloned toward the egress port. If there is no match, then it may undergo exception handling by the Table-miss flow entry; otherwise, it will be dropped.

The main components of a flow table are:

*Match fields:* ingress port and the patterns to be matched in the packet's header;

*Priority:* packet priority level;

*Counters:* data structures that are updated when a packet is matched;

*Instructions:* commands to modify the action set or the pipeline processing;

*Timeouts:* time before an entry expires;

*Cookie:* used by the controller to collect postmortem pipeline statistics.

A match in a table can trigger up to five actions:

1. Update the action set (i.e., specify how a packet is modified before exiting the switch);
2. Update the packet header (i.e., perform an immediate update of the packet's header information);

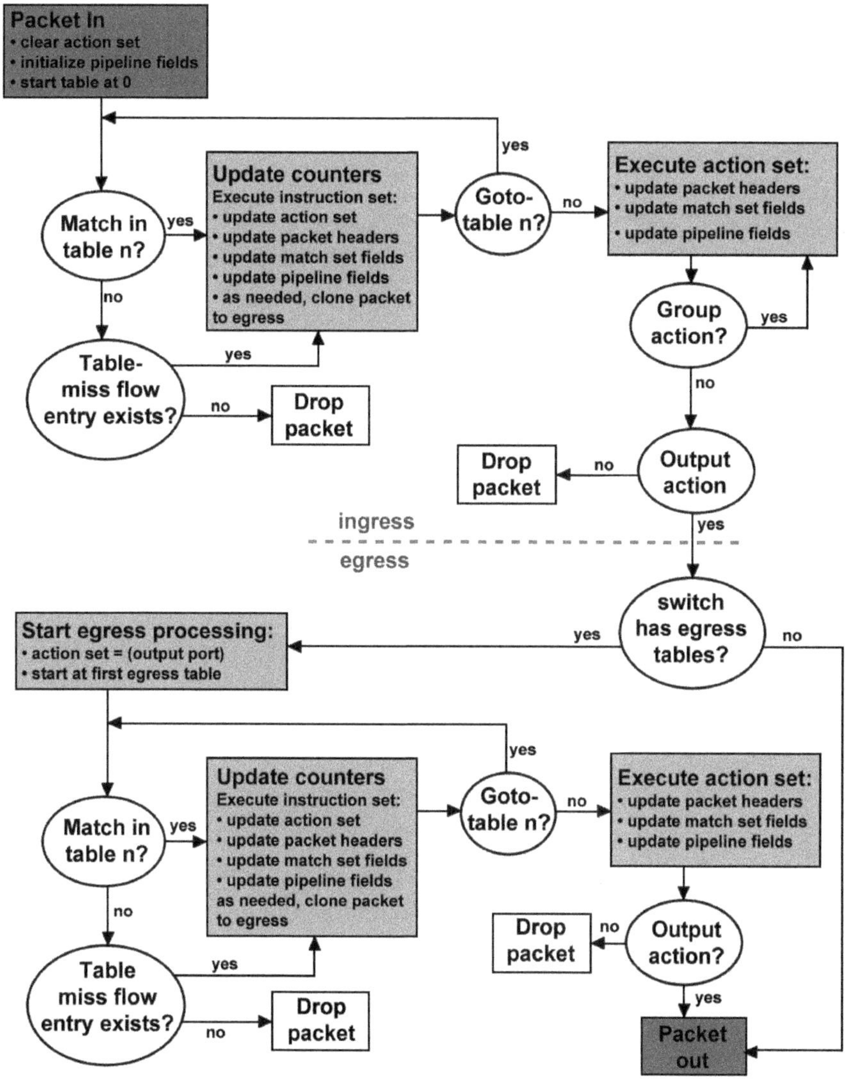


**Figure 4.4** Simplified flowchart of packet processing in an OpenFlow switch pipeline.

3. Update match set fields (i.e., changing matched header fields such as VLAN ids, MPLS label, or PBB values);
4. Update pipeline fields (i.e., packet registers);
5. Clone a packet and send it to an egress port.

### 4.2.7 Instructions and Actions

An important separation in the OpenFlow abstraction model is the differentiation between *instructions* and *actions*. The instructions describe the *processing of a packet with respect to the pipeline* and the actions *describe the changes on a packet and the final forwarding* to an egress port. The *Action Set* is associated with a packet and the actions in a set are accumulated during the flow. The actions are executed when the specific packet exits the pipeline.

A flow table entry contains one or more instructions that are executed when a packet matches the entry. The specification defines six different instructions, three required and three optional, which are

Required instructions:

*Write-Actions action(s):* Merges the specified action(s) into the current action set of the packet.

*Clear-Actions:* Immediately clears the action set.

*Goto-Table next-table-id:* Indicates the next-table-id in the processing pipeline.

Optional instructions:

*Stat-Trigger stat thresholds:* Generate an event to the controller if the statistics crosses a threshold.

*Apply-Actions action(s):* Applies the specified action(s) immediately and without any change to the Action Set. This action modifies the packet between tables or may execute multiple actions of the same type.

*Write-Metadata metadata/mask:* Writes metadata (typically using a mask) and is used to transfer state information between tables.

An Action Set is associated with each packet and is used to assemble actions to modify a packet. The specification's *required* and *optional* actions are

Required actions:

*Output:* Forward a packet to a specified egress port.

*Group group-id:* Send to a specific group in the list in the Group table.

*Drop:* The packet is dropped.

Optional actions:

*Meter meter_id:* Forward the packet to a specified meter. A packet might be dropped.

*Set-Queue:* Select a specific queue associated with a port. This provides the basis for QoS.

*Push-Tag/Pop-Tag:* Add (Push) or delete (Pop) tags from a packet, some switching protocols; for example, MPLS or different Ethernet versions (including VLANs and provider backbone bridging (PBB)), use this.

*Set-Field:* Alter the packet's header fields, also used for integration with existing networks (e.g., VLANs).

*Change-TTL:* Modification IPv4 TTL, IPv6 Hop Limit, or MPLS TTL values in a packet.

### 4.2.8 Header and Match Fields

OpenFlow's capability comes largely from the capability of simple flow handling. Flows are described by patterns that are matched with the information in the header fields. OpenFlow requires the matching of well specified headers. In addition to headers, table entries can match the physical and logical incoming port and metadata (i.e., state information that is passed on from one table to another). OpenFlow Version 1.5.1 defines the matching of 45 different header and match fields.

### 4.2.9 Examples for Matching Headers

One elegance of OpenFlow is the capability to manage advanced packet handling found in typical telecom networking while providing an abstraction that is more general. This enables a switch to be implemented with only the required functionality, possibly reducing costs.

The capabilities include typical data layer activities (including VLAN switching), network layer action (IP routing), and combined transport and application layer activities (e.g., firewalls or network address translation (NAT) functions).

Table 4.1 shows examples of these network functions on the data plane.

### 4.2.10 OpenFlow Protocol

The OpenFlow protocol specifies the communication between the controller and switch using secure communications. These are usually encrypted using TLS but may be run directly over TCP. OpenFlow uses a simple binary protocol with data structures defined by header files written in 'C' [8].

The protocol permits flexible binding between controllers and switches that assumes a one-to-one relationship of a controller with a switch (i.e., a switch is controlled by a single controller) but allows for multiple controllers to manage a switch (N:1 relationship); that is, N controllers might configure one switch so that different controllers are managing different network slices.

**Table 4.1**
Examples of OpenFlow Emulating Network Functions

| Emulated Activity | In Port | MAC src | MAC dest | Eth type | VLAN ID | IP Src | IP Dst | IP Proto | TCP src Port | TCP dst Port | Action |
|---|---|---|---|---|---|---|---|---|---|---|---|
| Switching | * | * | 00:1f:.. | * | * | * | * | * | * | * | port 6 |
| Flow switching | Port 3 | 00:20.. | 00:1f:.. | 0800 | Vlan1 | 1.2.3.4 | 5.6.7.8 | 4 | 17264 | 80 | Port 6 |
| Firewall | * | * | * | * | * | * | * | * | * | 22 | drop |
| Routing | * | * | * | * | * | * | 5.6.7.8 | * | * | * | Port 6 |
| VLAN switching | * | * | 00:1f:.. | * | * | * | * | * | * | * | Port 6, port 7, port 8 |

The protocol supports three main message categories: (1) the *controller-to-switch* messages that are initiated by the controller and used to directly manage or inspect the state of the switch (see Table 4.2), (2) *asynchronous* messages, sent without the controller soliciting them from a switch, these are used by the switches to denote a packet arrival or switch state change (see Table 4.3), and (3) *symmetric* messages that are sent in either direction without prior request (shown in Table 4.4).

The handling of packets outside a switch is an appealing feature of OpenFlow. The switch can use the Packet-in message to instruct the controller to act on the packet. OpenFlow allows for two packet-handling modes between switch and controller. First, if the switch has sufficient buffer availability, then the packet is temporarily stored in the switch and the switch transfers only the control and some packet information (e.g., the headers or fractions of them for

**Table 4.2**
OpenFlow Controller-to-Switch Messages

| Protocol Message Name | Semantics |
|---|---|
| Features | Query the switch to identify and report its capabilities. |
| Configuration | Queries and sets configuration parameters. |
| Modify-state | Manages the state of the ports. The primary purpose is to add, delete, or modify entries in flow tables, groups, and action buckets. |
| Read-state | Read from the switch (e.g., configuration, statistics, and capabilities). |
| Packet-out | Instructs the switch to send a packet out a specified port. |
| Barrier | Forces the switch to complete processing of certain messages before advancing (e.g., deleting all flow entries (flow-mod)). |
| Role-request | Sets the role of the OpenFlow Channel and the ID or queries it. |
| Asynchronous-configuration | Sets or queries the filter on asynchronous messages for a controller (used when multiple controllers applied). |

**Table 4.3**
OpenFlow Asynchronous Messages

| Protocol Message Name | Semantics |
|---|---|
| Packet-in | Transfer the control of a packet to the controller for further handling (typically a pointer but eventually including the whole packet). |
| Flow-removed | Informs the controller that an entry from a table is removed; a result of a controller's delete request or of an expiring flow. |
| Port-status | Used to notify a controller about changes of the ports' status or configurations. |
| Role-status | Informs the controller of a change of its role (e.g., a new controller becomes a new main controller for a switch). |
| Controller-status | Informs the controller about changes of the OpenFlow channel changes (e.g., for failover handling with controllers). |
| Flow-monitor | Informs the controller of a change in a flow table. A controller may define a set of monitors to track changes (not covered here). |

**Table 4.4**
Open Flow Symmetric Messages

| Protocol Message Name | Semantics |
|---|---|
| Hello | Exchanged between the switch and controller at startup. |
| Echo | Echo request/reply messages to verify the liveness of a controller-switch connection. |
| Error | Message to notify each other about problems. |
| Experimenter | Used to define and test new OpenFlow messages. |

treatment by the controller). In the second mode (i.e., when a switch runs out of buffer space or is simply not able to buffer a packet), it then sends the full packet to controllers.

### 4.2.11 Distributed Controllers and Flow Visor

An OpenFlow switch may be controlled by single or multiple controllers. The use of multiple controllers can increase the reliability of the system since a switch can continue to operate if one or more controllers fails or is compromised. In addition, the use of more than one controller can increase the performance of the control plane due to the split of work as well as allowing for the new concept of network slicing (i.e., for the controlled isolation of network resources while integrating the flows in a general substrate) (for more details on network slices see below). Figure 4.5 shows the example of a single controller managing multiple switches while Figure 4.6 shows the model of multiple controllers managing multiple switches. Network Sliced OpenFlow switch to controller options is shown in Figure 4.7.

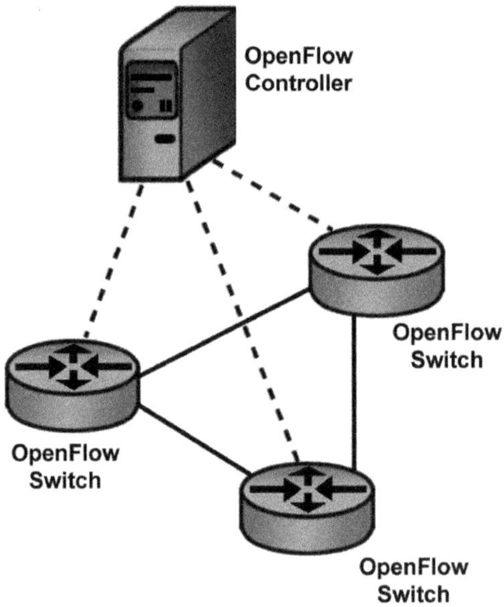


**Figure 4.5** 1:N controller to switch.

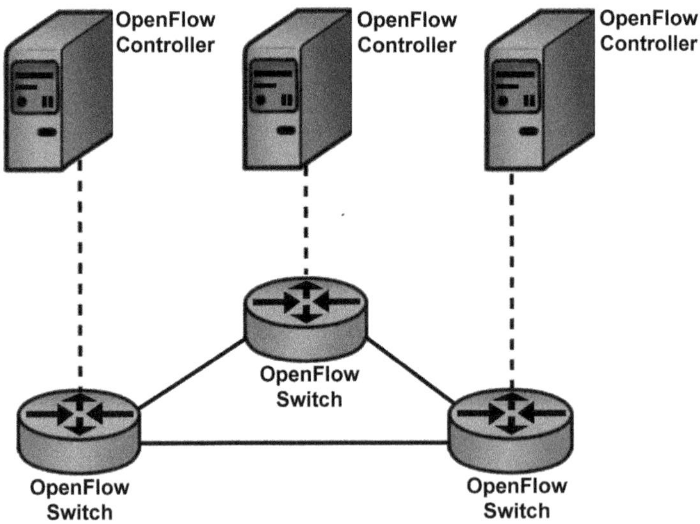


**Figure 4.6** N:M controller to switch.

The OpenFlow protocol provides mechanisms for coordination between controllers with different roles. An OpenFlow controller might have three different personalities: (1) it may be equal (i.e., the controller has full access to the switch, it receives all the asynchronous messages, and can send commands to

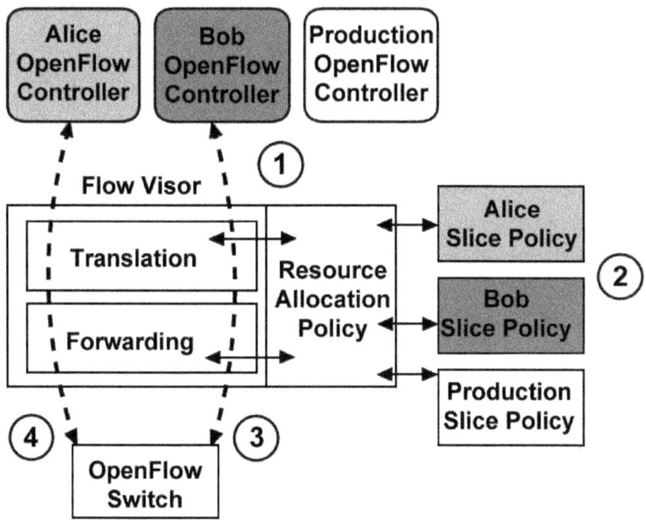


**Figure 4.7** Network sliced controller and switches.

modify the state of the switch, including adding or deleting flows), (2) it may be read-only (i.e., the controller has a read-only access to the switch and does not receive asynchronous messages apart from port status information), and finally, (3) a controller may be a control (i.e., it has full access to the switch as in the "equal" role); however, when the controller changes "control," the switch changes the controller previously in "control" to "read-only" role (controllers in the equal role are not affected by this change). (Legacy use of the terms have been altered to use "control" and "read-only.")

### 4.2.12 Evaluation of the OpenFlow Concept

OpenFlow is the first practical framework for implementing the concept of SDN to separate the data plane and control plane in fast Ethernet switches. Google announced in 2012 that it uses OpenFlow for traffic and flow management within its inter-datacenter networks [10]. OpenFlow can be credited with enabling SDN as a new networking breakthrough concept and for enabling truly programmable networks, yet the framework also contains major intentional deficiencies. First, the OpenFlow specification must be updated to accommodate changes or new networking protocols. OpenFlow also lacks an applicable programmability concept. Although the OpenFlow instruction set comprises a goto statement, it lacks the arbitrary definition of data structures and the use of more complex execution commands. These constrain the control and prevent parallel execution or "case" programming statements. Even with these limitations, commercial OpenFlow switches easily achieved the packet throughput required for telco networks.

### 4.2.13 The Importance of OpenFlow in 5G

Although switches and routers are rarely shown in system diagrams for the CSP network, they are abundantly deployed. The importance of the separation of the control plane from the data plane allows for the expansion of the control plane, when necessary, without impacting the data plane processing elements. Previously, upgrades in table space or processing power in the control plane would incur costly upgrades to controller cards in routers, or complete replacement of switches where the controller logic was integrated with the data plane. A second advantage is that switches and routers from different vendors no longer require different skill sets resulting from different command and operational protocols.

These advantages result in significant operational savings to the CSP.

It's now time to move on from OpenFlow to a relatively new option for building and controlling switching fabrics, Programming Protocol-Independent Packet Processors (P4).

## 4.3 P4

P4 provides a protocol and hardware independence to the programmability of flow management in switches and routers. First proposed in 2014 [11], it creates a domain-specific programming language (DSL) to create independence from, and separation between, hardware and software, which sounds like virtualization. It is built around the separation of data plane and control plane as well as the localization of control action by exploiting the fast-forwarding (acceleration) capabilities of switches. P4 addresses the disadvantages in OpenFlow with a more expandable forwarding model using greater generalization. This model allows for an easier adaptation to new hardware capabilities and new network protocols and their packet formats. The expandability provides for the programmability of packet handling beyond the serialization done by pipelines (i.e., eventually the specification of parallelism in packet processing). As a result, P4 might even permit faster packet forwarding by programmability of the switch fabric.

P4's abstractions and forwarding model generalizes how packets are processed by different forwarding devices, mainly Ethernet switches, but it is also suitable for load balancers or routers based on different technologies, such as ASICs (and network processing units (NPUs), which are similar to CPUs, but specific to support acceleration of network workloads; also called infrastructure processing units (IPUs) by one vendor), software switches, or FPGAs. This allows P4 to use a common domain-specific programming language to represent how packets are processed in view of its abstract model. Consequently, developers and operators of P4-based SDN networks can create target-independent programs that a compiler can map to a variety of different forwarding devices. These devices can range from relatively slow software switches to the fastest ASIC-based switches. In fact, there are P4-specific hardware devices in the market already.

P4's major mechanism is the use of the P4 programming language. In contrast to general-purpose languages such as C or Python, P4 has several language constructs optimized for forwarding. P4 is distributed as open-source, it is permissively licensed code, and it is maintained by the P4 Language Consortium, a not-for-profit organization hosted by the ONF. The current revision of the P4 language can be found at [12]. The P4Runtime API is a control plane specification to manage the data plane elements of a device defined by a P4 program [13].

### 4.3.1 Domain-Specific Programmability

Domain-specific programming languages have recently become popular due to the rise of domain-specific modeling (DSM) [14]. DSM is a software engineering concept that derives computer software from a model rather than writing the code itself. A DSM-based model is characterized by the systematic use of a description language to represent the system's various features. An important capability of programming languages is the one denoted as Turing completeness (i.e., they can be used to simulate any Turing machine). It was recently shown [15] that remote direct memory access (RDMA) NICs are Turing complete and capable of performing arbitrary computation. This capability allows one to use them as smart forwarding elements.

### 4.3.2 The P4 Language

The P4 language is a structured programming language (i.e., there is no jump/goto). P4 avoids the well-known deficiencies of nonstructured languages (see Dijkstra's seminal work on nonstructured programming languages in [16]). P4 also supports features and capabilities of functional and procedural programming languages, like C, C++, Java, and Python. This enables modularity and code reuse. The domain-specific language concepts in P4 are

- The use of commands, declarations, and variables to describe the packet forwarding rules instead of just configuring the tables with patterns and actions;
- The use of a hardware-specific P4 runtime environment that is split between SDN controllers and switches, the messages and the semantics

for the interfaces between the controllers (client(s)), and the switch (here the server) are well defined;

- The use of a compiler to translate P4 programs into P4 API calls.

P4 programs execute a constant number of operations for each byte of an input packet. Although parsers may contain loops, provided some header is extracted on each cycle, the packet itself provides a bound on the total execution of the parser. In other words, the computational complexity of a P4 program is linear in relation to the size of the headers and never depends on the size of the accumulated state while processing data. This guarantees fast packet processing across a variety of targets.

### 4.3.3 P4 Concept

The initial concept of P4 was to complement OpenFlow's capabilities to describe and configure packet-forwarding in fast switches. P4 borrowed the concepts of local flow switching and management from OpenFlow as well as the separation in SDN between the control plane and the data plane, as shown in Figure 4.8. This graphic depicts the classical OpenFlow process that the SDN controller populates the flow tables and rules on the target switch. The switch itself translates the rules into actions. P4 adds to this process the ability to specify a *P4 program* that describes how the switch should process and forward the packets. This program is parsed by a compiler and translated into rules and code for the switch, shown on the left side in Figure 4.8.

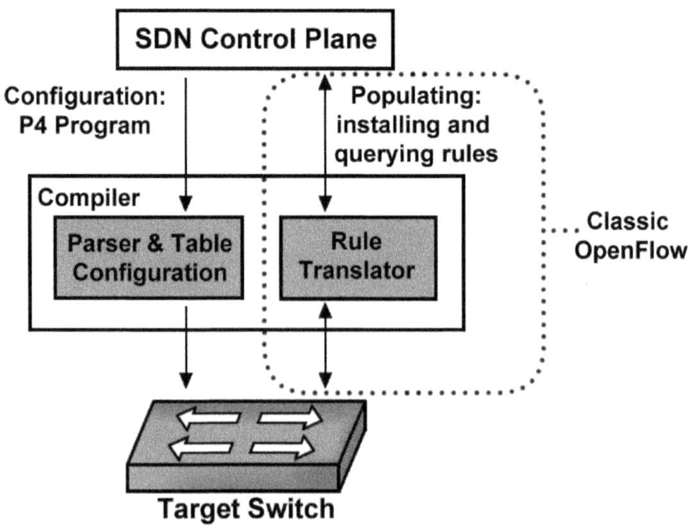


**Figure 4.8** P4 programmability concept vs population and configuration in OpenFlow.

### 4.3.4 Data Plane Forwarding and P4 Enhancements

P4 uses an enhanced generalization of OpenFlow's switch model for forwarding. This generalization has become known as the Portable Switch Architecture (PSA), and has been evolved over time by various teams in academia and industry [17–19]. In the P4 model, switches forward packets after they are processed by the programmable parser, shown in Figure 4.9. After the parser, the packet is handled by multiple stages of match + action processing activities. These activities can be arranged in a sequence, in parallel, or any combination of both.

P4 adds three generalizations to the forwarding model of OpenFlow:

1. P4 supports a programmable parser to allow new headers to be defined;
2. P4 allows for parallel packet pipelines (i.e., packets can be processed by a sequence of match + action activities or parallel ones);
3. P4 assumes that actions are composed from protocol-independent primitives and can be supported by the switch (e.g., the construction of lookup keys from packet fields defined at runtime or from computed metadata).

P4's forwarding model is embedded in the P4 runtime environment controlled by "Configure" and "Populate" operations. The *Configure* operations determine which protocols are supported and how a P4 switch processes the packets. These operations program the parser, define the pipelines (i.e., set the order of the match + action stage), and specify the header fields processed by

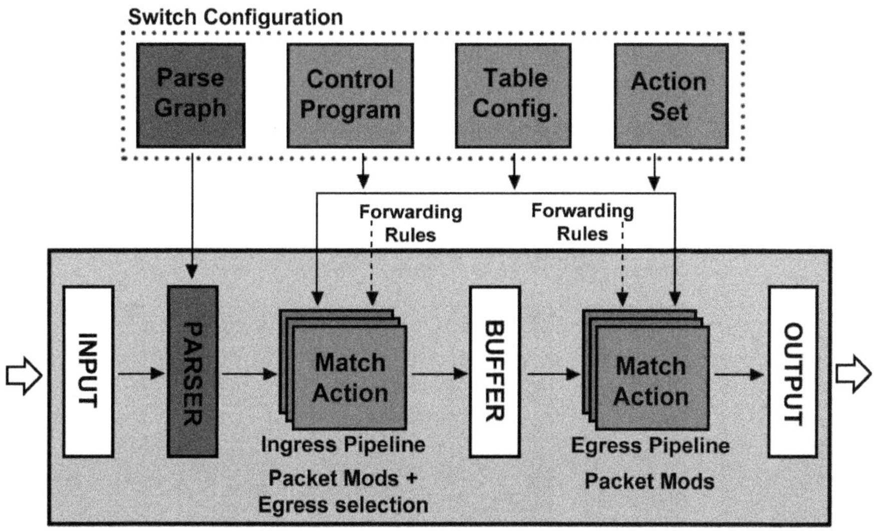


**Figure 4.9** Switch configuration and forwarding in P4.

each stage. The *Population* operations add and remove entries to the tables. They determine the policies applied to packets at any given time.

### 4.3.5 Portable Switch Architecture

Figure 4.10 shows a simplification of P4's Portable Switch Architecture to aid in understanding the handling of the packet headers. Arriving packets are first handled by a programmable parser.

The body of the packet is buffered as shown in Figure 4.11 and not available for matching. The parser recognizes and extracts header information from the packet. P4 does not make any assumption about the meaning of the header fields, and thus, what protocols are supported. The extracted header fields are then handed to the match + action tables. The tables' match + action activities may modify the packet headers and inject meta information (i.e., intermediate results and information associated with the header). P4 also allows adding headers. At the end of the programmable match + action pipeline the packet headers are deparsed (i.e., they are rebuilt for a packet and serialized for egress treatment of the packet or further packet processing).

### 4.3.6 Programming a P4 Device

The P4 design centered around the programmability of flow handling. P4 intertwines the aims with the structure of a protocol-independent and portable switch architecture. The approach led to a simple but elegantly structured programming workflow for developers to create P4 programs. This workflow comprises three main phases when coding the flow handling in P4:

1. A programmer declares the headers that should be recognized in the flow handling and their position in the packet. This step is required since the protocols are a priori not defined and the programmer is given the freedom to define and adapt to protocol changes.

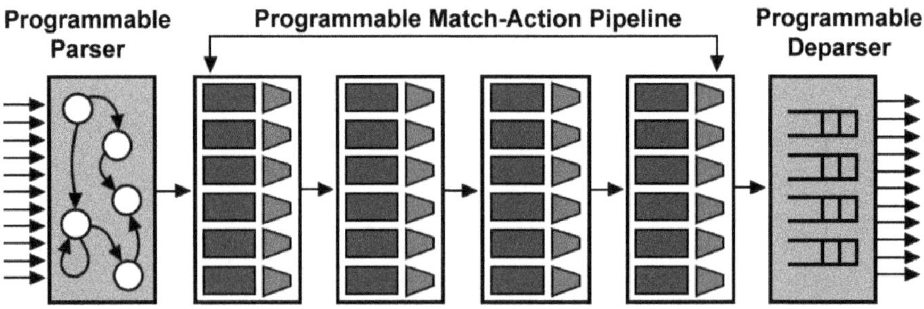


**Figure 4.10** P4 match + action pipeline with programmable header parser and deparser.

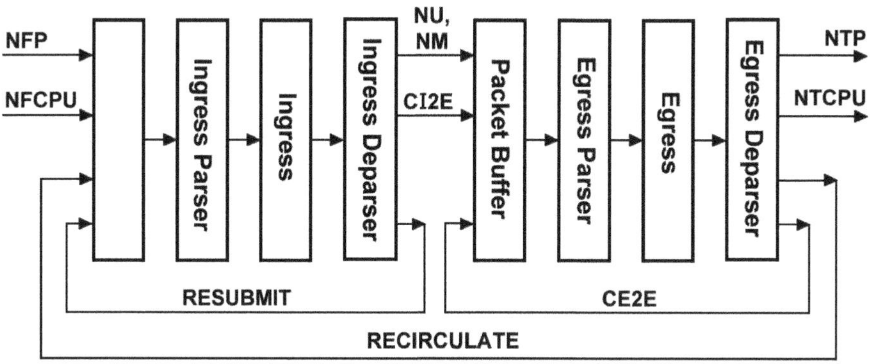


**Figure 4.11** Packet path in P4's Portable Switch Architecture.

2. The programmer defines the tables and the exact processing algorithm. This is the actual behavior of the packet processing for the flow handling.
3. Finally, the programmer declares how the output packet will look on the wire. This focuses on the composition of the new header of a packet.

The application of the phased approach in P4 programming is supported by a set of programming tools and constructs similar to other programming tools. These tools and components comprise a P4 program, a P4 compiler, and P4 APIs. A typical tool workflow when programming a target device using these components is shown Figure 4.12.

A manufacturer of a switch or a target device provides the hardware or software architecture definition and a P4 compiler for that target. P4 programmers write programs for a specific architecture that defines a set of P4-programmable components on the target as well as their external data plane interfaces. When a programmer compiles a P4 program the compiler generates two artifacts:

1. A data plane configuration that implements the forwarding logic described in the P4 program;
2. An API for managing the state of the data plane objects from the control plane.

P4 programs consists of the four components shown in Figure 4.13. Two are the rarely updated components: the P4 language specification itself, and the core.p4 library that declares some built-in P4 constructs. Furthermore, the manufacturer provides architecture-specific software: the P4 description of the target's architecture (arch.p4), and the library of P4 API calls for this architecture

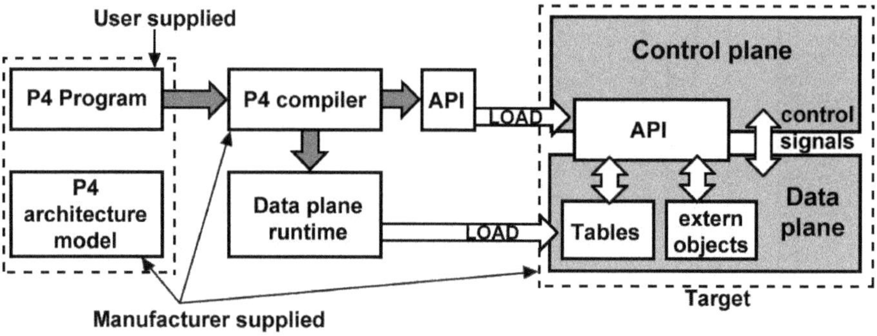


**Figure 4.12** Programmability tools and components in a P4 environment.

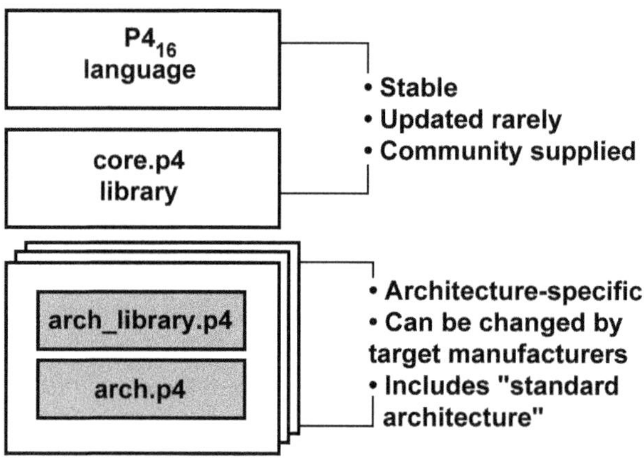


**Figure 4.13** Environment for P4 programs.

(arch_library.p4). These software components can be updated when the target hardware or software changes over time.

### 4.3.7 The P4 Language

Instead of describing and providing a formal description of the P4 language, we will describe the P4 language with a programming example. This follows P4's fundamental concept of enabling flow handling by using already well-known and understood general programmability mechanisms. We base our description on excellent tutorials provided by the P4 consortium at P4.org [20].

Tables 4.5 and 4.6 depict a typical template for a P4 program. At first, the program includes the general definitions given in the core.p4 file as well as the target's architecture definition, denoted here as v1model.p4. The architecture

**Table 4.5**
P4 Program Template

```
#include <core.p4>
#include <vlmodel.p4>
// headers
struct metadata { ... }
struct headers {
  ethernet_t ethernet'
  ipv4_t     ipv4;
}

// Parser
parser MyParser ( packet_in packet,
                  out headers hdr,
                  inout metadata meta,
                  inout standard_metadata_t smeta
                ) {
                ...
}

.. checksum Verification
control MyVerifyChecksum ( in headers,
                           inout metadata meta
                         ) {
                         ...
}

// Ingress Processing
control MyIngerss ( inout headers hdr,
                    inout metadata meta,
                    inout standard_metadata_t std_meta
                  ) {
                  ...
}

// Egress Processing
control MyEgress ( inout headers hdr,
                   inout metadata meta,
                   inout standard_metadata_t std_meta
                 ) {
                 ...
}

// Checksum update
control MyComputerChecksum ( inout headers hdr,
                             inout metadata meta
                           ) {
                           ...
}
```

**Table 4.5** (continued)

```
// Deparser
control MyDeparser ( inout headers hdr,
                     inout metadata meta
                   ) {
                   ...
}

// Switch
V1Switch (
  MyParser (),
  MyVerifyChecksum (),
  MyIngress (),
  MyEgress (),
  MyComuterChecksum (),
  MyDeparser ()
) main;
```

definition is typically supplied by the manufacturer. The switch is named V1Switch and its packet handling is defined by the main function. The main function executes a sequence of commands for packet and flow handling. This handling comprises the programming of the header parser by the MyParser() function that is followed by calling the MyVerifyCheckSum(), which verifies the header extraction. After the correct extraction, the MyIngress() function specifies the packet handling in the ingress pipeline. Succeeding this step, the packet is handed over to the egress pipeline. The packet handling for the egress pipeline is defined by the MyEgress() function. Next in the sequence is verification that the information for a new packet header is correct, and finally, the rebuilding of the new packet header occurs and is specified by the MyDeparser() function.

Table 4.6 shows a very simple example for packet forwarding programmed in P4. Here, all packets from port 1 of the switch are forwarded to port 2 and vice versa. The parser does not need to extract any headers from a packet and defines only metadata. The packet is passed along to the MyVerifyChecksum() function block, which also doesn't execute anything (since headers are not considered). In the next step, the actual forwarding occurs. When the packet arrives at the MyIngress() block, the P4 program checks the port on which the packet arrived by reading the ingress_port field of the standard_metadata information and decides which port to send the packet to using a simple if, else-if statement. The remaining function blocks (MyEgress, MyComputeCheckSum, MyDeparser) do not need to do anything and at the end of the program we instantiate the switch. In this example, the connection between port 1 and port 2 is hardcoded and cannot be changed after the P4 program is compiled.

**Table 4.6**
A Simple P4 Program for Packet Forwarding

```
    #include <core.p4>
    #include <v1model.p4>
    struct metadata {}
    struct headers {}

    paarser MyParser (packet_in packet,
                      out headers hdr,
                      inout metadata meta,
                        inout standard_metadata_t standard_meta-
data) {

                      state start {transition accept; }
    }

    control MyVerifyChecksum (inoiut headers hdr, inouit meta-
data meta ) { apply { } }

    control MyIngerss (inout headers hdr,
                       inout metadata meta,
                      inout standard_metadata_t standard_metadaa
) {

                       apply {
                             if (standard_metadata.ingerss_port
== 1) {
                             standard_metadata.egerss_spec = 2;
                               } else if (standard_metadata.in-
gerss_port == 2 ) {
                             standared_metadata.egress_spec = 1;
                                 }
                              }
                        }

    control MyEgress (inout headers hdr,
                      inout metadata meta,
                        inout standard_metadata_t standard_meta-
data ) {

                        apply  { }
                      }
    control MyComputeChecksum (inout headers hdr, inout metadata
meta ) {
                        apply { }
                      }
    V1Switch (
      MyParser (),
      MyVerifyiChecksum (),
      MyIngress (),
      MyEgress (),
      MyComputeChecksum (),
      MyDeparser ()
    ) main;
```

### 4.3.8 P4 Runtime Architecture

Figure 4.14 details the P4 Runtime Reference Architecture. The runtime environment allows the control and executing of the packet forwarding. The environment encompasses the controller (shown at the top of Figure 4.14), and the target device (shown at the bottom of Figure 4.14). The controller operates as a client and the target is the server. P4 allows for multiple controllers but grants only write access to a single primary controller for each read/write entity. P4 uses a client arbitration scheme that ensures that only one controller has write-access to each read/write entity or to the pipeline configuration itself. Any controller may perform read access to any entity or the pipeline configuration.

An important part of P4 is the P4Runtime API. The API defines the messages and semantics of the interface between the client(s) (i.e., controllers) and the server (i.e., the controlled target switch). The API is specified by the p4runtime.proto file, which may be compiled to produce both client and server implementation stubs in a variety of languages. The controller can access the P4 entities that are declared in the P4Info metadata. The P4Info structure is defined by p4info.proto, another specification file available as part of the P4 standard.

### 4.3.9 Evaluation of P4

P4 has significant advantages over OpenFlow or other packet-processing systems (e.g., the ones that are built mainly around microcode on top of custom hardware). These advantages are:

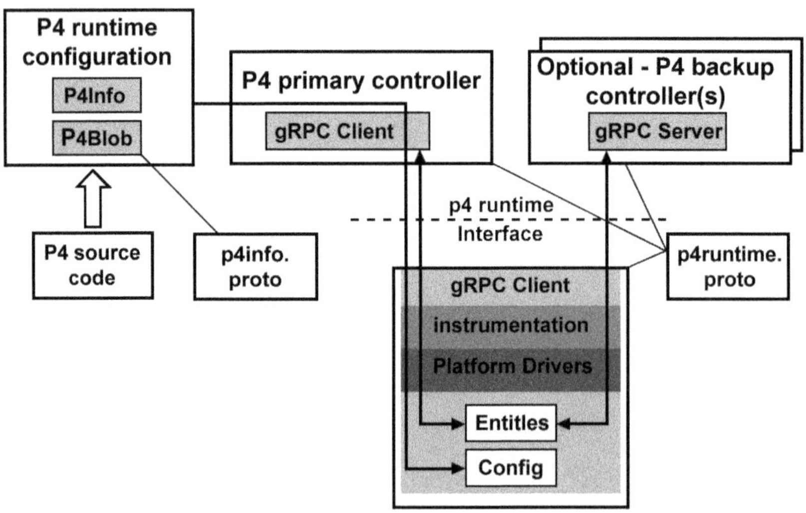


**Figure 4.14** The P4 Runtime Architecture.

- *Flexibility:* Allows packet-forwarding policies to be expressed as programs.
- *Expressiveness:* Permits hardware-independent packet processing algorithms that apply general-purpose operations and table look-ups. P4 programs are portable across hardware targets using compilers that implement the same architectures.
- *Resource mapping and management:* Programs describe storage resources abstractly (e.g., IPv4 source address), and the compiler maps user-defined fields to available hardware resources and manages low-level details such as allocation and scheduling.
- *Software engineering:* Type checking, information hiding, and software reuse are supported.
- *Component libraries:* Component libraries supplied by manufacturers can be used to wrap hardware-specific functions into portable high-level P4 constructs.
- *Decoupling hardware and software evolution:* Target manufacturers may use abstract architectures to further decouple the evolution of low-level architectural details from high-level processing.
- *Debugging:* Manufacturers can provide software models of an architecture to aid in the development and debugging of P4 programs. The P4 community and independent developers can write programs that can verify the P4 program and enable in-operation error localization.

The current state-of-the-art silicon that supports P4 is capable of over 12.8Tb/S switching on 400Gb/S ports.

## 4.4 Conclusion

SDN-capable switches have been deployed in abundance in today's 5G system. The advantages of traffic handling by programmable flow management, as facilitated by OpenFlow and P4, may yield significant operational savings to a CSP. However, there are additional options to accelerate and virtualize the data plane; some of these will be covered in later chapters.

## References

[1] Clark, D., "The Design Philosophy of the DARPA Internet Protocols," in *Symposium Proceedings on Communications Architectures and Protocols (SIGCOMM'88),* 1988, pp. 106–114.

[2] Wikipedia: Procedural Programming, https://en.wikipedia.org/wiki/Procedural_programming.

[3] Wikipedia: Functional Programming, https://en.wikipedia.org/wiki/Functional_programming.

[4] Rommer, S., et al. *5G Core Networks: Powering Digitalization,* London: Academic Press, 2019.

[5] European Telecommunications Standards Institute/3GPP: *TS 123 501–System Architecture for the 5G System (5GS), Version 16.10.0 Release 16,* September 2021.

[6] Caesar, M., et al., "Design and Implementation of a Routing Control Platform, in *Proceedings of the 2nd Conference on Symposium on Networked Systems Design & Implementation (NSDI05),* Vol. 2, 2005, pp. 15–28.

[7] European Telecommunications Standards Institute/3GPP: *TS 23.214–Technical Specification Group Services and System Aspects; Architecture Enhancements for Control and User Plane Separation of EPC Nodes; Stage 2 (Release 17),* June 2021.

[8] McKeown, N., et al., "OpenFlow: Enabling Innovation in Campus Networks," *ACM SIGCOMM Computer Communication Review,* Vol. 38, No. 2, 2008, pp. 69–74.

[9] Nygren, A., et al. "OpenFlow Switch Specification Version 1.5. 1," Open Networking Foundation, Technical Report, 2015.

[10] Levy, S., "Going with the Flow: Google's Secret Switch to the Next Wave of Networking," *Wired,* 2012, https://www.wired.com/2012/04/going-with-the-flow-google/.

[11] Bosshart, P., et al., "P4: Programming Protocol-Independent Packet Processors," *ACM SIGCOMM Computer Communication Review,* Vol. 44, No. 3, 2014, pp. 87–95.

[12] The P4 Language Consortium, $P4_{16}$ Language Specification (Version 1.2.2), May 2021.

[13] P4.org API Working Group: P4Runtime Specification, Version 1.3.0, July 2021.

[14] Kelly, S., and J.- P. Tolvanen, *Domain-Specific Modeling: Enabling Full Code Generation,* Hoboken, NJ: John Wiley & Sons, March 2008.

[15] Reda, W., et al. "RDMA is Turing Complete, We Just Did Not Know It Yet!" arXiv preprint, 2021, arXiv:2103.13351.

[16] Dijkstra, E. W., "Letters to the Editor: Go To Statement Considered Harmful," *Communications of the ACM,* Vol. 11, No. 3, 1968, pp. 147–148.

[17] Barefoot, Tofino. https://www.barefootnetworks.com/products/brief-tofino-2/.

[18] Bosshart, P., et al., "Forwarding Metamorphosis: Fast Programmable Match-Action Processing in Hardware for SDN," *SIGCOMM Comput. Commun. Rev.,* Vol. 43, No. 4, August 2013, pp. 99–110.

[19] P4.org Architecture Working Group, P416 Portable Switch Architecture (PSA), https://p4.org/p4- spec/docs/PSA.html.

[20] Ibanez, S., B. O'Connor, and M. Arashloo, P4 Language Tutorial, https://bit.ly/p4d2-2018-spring, 2018.

# 5

# Performance of Infrastructures for Virtual Network Functions

## 5.1 Performance and Security Considerations

Performance is an important aspect referred to as a *nonfunctional (performance and structural) attribute of ICT systems.* Such attributes describe characteristics of reliability, performance, maintainability, scalability, and usability of systems. They characterize operational capabilities of systems, in contrast with the functional features whose aim is to accomplish the computational or communication tasks of ICT systems.

The requirements for nonfunctional and operational features for ICT systems are often addressed by choosing from a number of interchangeable technologies that individually achieve the same main functionalities. Choosing which technology should be applied is typically not trivial; rather it is an important design decision that impacts the operational features of technical systems such as their OpEx.

The trade-offs between the features that are provided by these technologies (e.g., usability versus security) makes design decisions challenging. Optimizing features that improve the nonfunctional attributes may lead to a complex design process and eventually increases design and acquisition costs. The balancing of the trade-offs to achieve schedule and development cost targets may have a very significant impact on the final OpEx for the operator once the product is deployed, or may render the product brittle or too costly to deploy. A system designer needs to understand how to evaluate the different technologies that are available and choose trade-offs wisely.

This chapter discusses the fundamentals of performance evaluation in virtualized software-defined infrastructures and networks with specific emphasis on cloudified RAN. The requirements for virtual infrastructures on performance are discussed first. The chapter then describes basic models for performance, and then discusses initial approaches to take advantage of design trade-offs. Special consideration is given to the virtualization modes and resource-sharing concepts in virtualized mobile core and RAN infrastructures.

### 5.1.1 Virtualization Modes and Requirements

The term "virtualization" in its widely accepted definition in ICT systems is the act of creating a virtual (rather than actual) version of an ICT object such as virtual computer hardware platforms, storage devices, or network resources. This definition is aligned with the three main requirements for virtualized systems as outlined in Chapter 3. These requirements postulate that virtual systems must not differ in the functional capabilities from their physical equivalents. They need to be as indistinguishable as possible when it comes to system operation (e.g., typically the operational overhead should be minimal).

The introduced overhead when managing virtual resources is straightforward to calculate, for example, the additional processing necessary for task switching in a CPU and any associated networking delay within a machine. However, often the practical overhead is not that obvious, such as how much more CPU performance is needed to run a virtual machine with the same responsiveness as a purpose-built system. The next sections highlight these issues and present initial and general models for assessing the overhead and other similar burdens incurred by virtualized systems.

### 5.1.2 Sharing, Aggregation, and Emulation in Virtualization

Three basic modes of operation exist when virtualizing computer and communication systems and mimicking physical resources. These modes are depicted in Figure 5.1.

The most well-known virtualization method is the *sharing mode*, as seen in Figure 5.1(a). This is also known as multiplexing in some discussions by other authors. The sharing mode exposes a *single* physical resource to be perceived as *multiple* resources. In sharing mode, a physical resource such as a CPU core or memory is considered as *atomic* (i.e., it cannot be divided into further parts and its access is granted exclusively to a single request). Hence, the resource can only be accessed by a single request at a time.

The *isolation of resources* in the sharing mode is usually obtained by *indirection*. In general, indirection replaces the direct connection between a variable

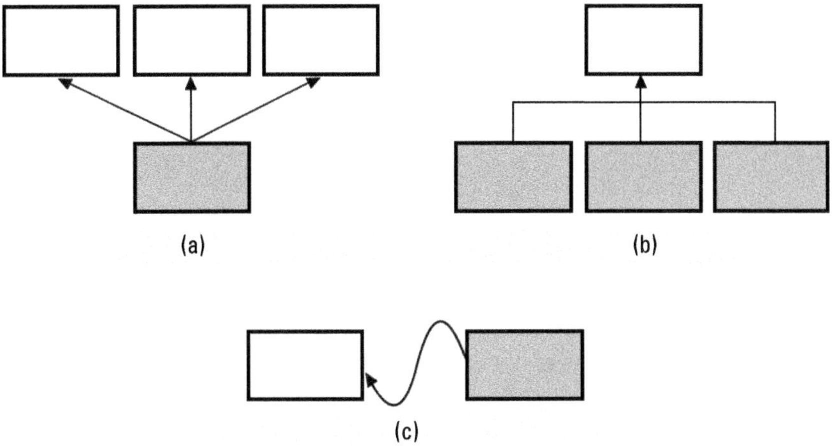


**Figure 5.1** Different virtualization modes: (a) sharing, (b) aggregation, and (c) emulation.

and its value, $v \rightarrow x$, with an indirect connection or link, $v \rightarrow u \rightarrow x$ [1]. Resource indirection is the ability to refer to a resource using a name, reference, or container instead of using the physical or genuine address of the object. The operation of indirection requires an *adapter* or a *shim*, which is an entity that is able to intercept a request, translate (i.e., to identify and to look up the translation), and redirect the request to a genuine and unique physical resource. The indirection in computer system virtualization is typically accomplished by the *hypervisor*, which is also referred to as a *virtual machine monitor* by some authors.

Today, the time-sharing mode in virtualization is assisted by special hardware features for improved performance. For example, modern microprocessors, such as Intel's Core™ and Xeon™ or AMD's Ryzen™ CPUs, have been enhanced during the last decade to include on-chip hardware logic to achieve fast task switching and fast memory management such as input–output memory management units (IOMMUs). These virtualization technologies are branded as VT-x by Intel and AMD-V by AMD.

Another virtualization method is the *aggregation mode*, shown in Figure 5.1(b). This mode bundles multiple physical resources and exposes them as a single resource. A resource object that is obtained through virtualization by aggregation is not atomic and can accommodate multiple requests at the same time. The aggregation mode often leads to enhanced capabilities of the resource. The parallel execution of the requests permits an increase in performance and throughput. The throughput scales almost linearly with the number of physical resources bundled into the single virtual one. The scaling, however, might be constrained eventually by granularity effects or the scheduling overhead when assigning atomic requests to the individual physical resources. Another advantage offered by the aggregation mode is the improved reliability of virtualized resources. For example, the hardware failure of an atomic resource within an aggregated bundle can be overcome by seamlessly reassigning the request to a different physical resource that is still operational. Ideally, the occurrence of a hardware failure has no or only minimal impact on the functionality and/or performance of the request under execution.

Typical examples of virtualized resources in aggregation mode include the redundant array of independent disks (RAID) storage systems in computer systems and the bundling of ports and links—obtained using the Link Aggregation Control Protocol (LACP)—in the Ethernet network standard.

The aggregation mode is of particular interest in cloud computing. A cloud system bundles many computational resources together and makes them accessible to users through network connections. The aggregation mode is applied in cloud computing to computing, storage, and software resources at once. By providing their resources elastically from a common resource pool, cloud computing achieves on-demand 24/7 service availability.

Finally, the third major virtualization method is through *emulation,* shown in Figure 5.1(c). This concept uses software to mimic the behavior of a real or hypothetical resource. The virtual resource behaves as required, like a real resource would, *but without requiring that such a physical resource exists.*

The major advantage of emulation is that the virtual resource may have features that do not exist yet in any real-world instantiation. Such features, for example, are hardware functionalities that are expensive to build but which must be tested before they are physically implemented. Another example is hardware that has been deprecated or is no longer in existence.

Emulation is typically achieved by executing a program that provides a service through an API and which is executed by a general computing entity. The *isolation of emulated resources* is typically achieved by provisioning them as a service. Emulated resources have well-defined programming (APIs) or service interfaces through which a user can access them. These service access points can be used as policy enforcement points, where users are required to be authenticated and authorized. Safe and secure execution of an emulated entity is achieved through isolated and trusted computing environments. A critical task of these environments is to protect the system itself (e.g., unintended or intended harmful operations that are applied to a virtualized function must be controlled and swiftly stopped). A relatively new requirement on execution environments comes from the large number of stakeholders now reliant on these environments. Arbitrary software can be provided from the outside by different parties and the software needs to be protected against exploitation even by the operators of the execution environment. A new trend to secure the code in such environments and to make the code inaccessible even by operators is called *confidential cloud computing* [2].

## 5.2 Performance Evaluation Concepts for the Sharing of Resources

A fundamental requirement in virtualization is to achieve a level of performance when using virtual resources that is indistinguishable from that obtainable using conventional physical resources. As a result, the functional performance metrics for both virtual and physical resources must be the same. These metrics are, for example, response time, correctness of computing, reliability, and probability of successful completion of tasks.

However, providing a virtual resource requires extra management and coordination. These tasks typically introduce some level of overhead. Not surprisingly, the efficiency of the virtualization mechanisms and the quality of their coordination are the main nonfunctional performance characterizations when using virtual resources. Any substantial performance deviation (reduction) that might be experienced when resorting to virtual resources—when compared to the case of physical resources—may hinder the applicability of the virtual resource.

This section details how to assess, using mathematics, the performance deviation caused by virtual resources. The concept is illustrated using a computer programmed to act as multiple routers. The example predates and is independent of the availability of hardware-support for network virtualization and SDN, such as the DPDK [3] or P4 (see Chapter 4). However, it still provides a number of performance parameters that may be adopted to assess the impact of coordination when using sharing mode in a networking scenario, providing fundamental insights into the performance assessment.

### 5.2.1 Networking Scenario

This scenario assumes that users are offered their own virtual router on physical hardware (bare-metal), as shown in Figure 5.2. The virtual routers running in VMs are provisioned through a virtualization environment that manages the shared resources (e.g., CPUs or RAM in the router and/or bandwidth of the attached physical links).

The virtual router's task is to perform the route decision and to forward packets. The routing function, specifically the control plane functionality, is reduced to a corner case in which the router must identify and match a packet from an ingress port while the egress port is always the same and predetermined. The data plane function of the virtual router is the forwarding of the packet(s) from the ingress port to the egress port.

Sharing mode requires that virtual routers share various physical resources of the physical router like CPU cores, RAM, and bandwidth of the single outgoing link in our example, including the egress port for every packet of every virtual router. Furthermore, all virtual routers in this scenario share the same

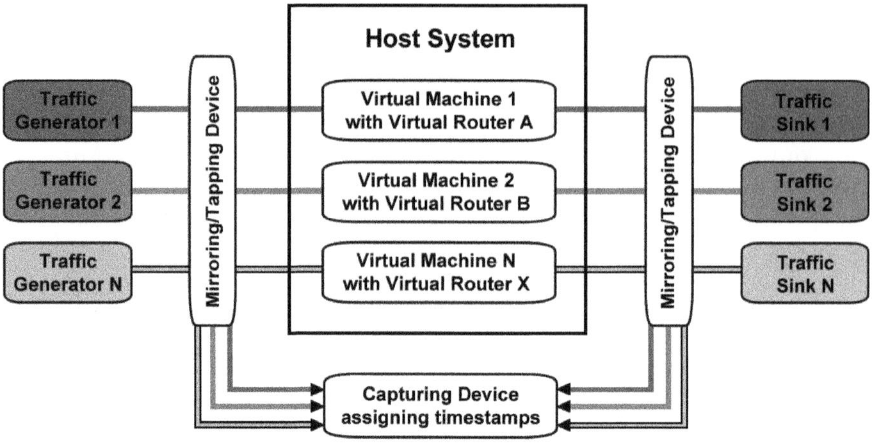


**Figure 5.2** Networking scenario for evaluation of shared mode.

physical input link (i.e., the ingress ports). The scenario assumes that the incoming traffic is well-constructed to avoid disturbances or contention on the physical ingress port (e.g., a packet generator for each virtual router feeds the router with real-world packets). The packet generators are programmed to generate repetitive patterns and are coordinated with one another to produce actual packets that do not cause contention on the ingress link. Every virtual router must match the incoming packets and forward them appropriately.

The scenario under consideration is equipped with mechanisms and hardware that can observe packet arrivals without altering the packet stream. The observation hardware records both the arrival time of each packet at the physical ingress port and the departure time of each packet at the egress port.

The hardware and software for the forwarding and the control plane of this router is a standard PC platform using an x86 CPU architecture, a standard PCIe bus, and standard Ethernet PCIe network interface cards for the physical ports. Each virtual router is instantiated by a full virtual machine that is managed by a virtualization environment like VMware Workstation [4] or VirtualBox [5] and executes a Linux distribution that runs routing and forwarding software (e.g., Iptables [6, 7]).

### 5.2.2 Mathematical Concept

The mathematics used to estimate the performance of virtualization mechanisms when in sharing mode is based on the concepts of perturbance analysis and jitter evaluation. As previously discussed, the virtualization mechanism requires a redirection layer that ideally should be transparent, but in practice affects the resources performance as it is perceived by the virtualized application, in this case the virtual router.

In more detail, the analysis presented here is based on comparative perturbance analysis. This concept was originally developed to characterize the network's impact on observable performance [8], like the change of packet interarrival times caused by a router. The mathematics of this analysis was later refined to characterize the quality of virtualization technologies (discussed next) [9] and estimate the quality-of-experience (QoE) faced by a user or an application based on the QoS, which is observed at the network level [10].

The main point of the suggested method is to compare the perturbance by virtualization with the performance of the activity when not using virtualization. This approach relates an observation (e.g., QoS) for a packet stream before (ingress) entering a bottleneck (i.e., virtual object) with the QoS object for stream after leaving bottleneck (egress). The comparison operation provides the name for the method. The analysis presented here enhances conventional jitter investigations [11] by explicitly considering the performance impacts on different timescales.

### 5.2.3 Mathematics Model

The following equations provide enhanced jitter analysis of packet once processed by the virtual router. The overall model and statistical objects are depicted in Figure 5.3.

Let $T_k^{in,i}$ (5.1) be a random variable (RV) defined as the interarrival time between two consecutive packets at the ingress port. The indices $i$ denote the virtual router number in this example and the index $k$ the sequence number of the packet in considered stream. Similarly, let $T_k^{out,i}$ (5.2) be the interarrival time between two consecutive packets at the egress port for the virtual router and the packet stream. The shared resources on the host system are considered atomic, for example a link, the bus, or CPU in the server. Hence, the RVs for the interarrival times are

$$T_k^{in,i} = t_k^{in,i} - t_k^{in,i-1} \tag{5.1}$$

$$T_k^{out,i} = t_k^{out,i} + t_k^{out,i-1} \tag{5.2}$$

#### 5.2.3.1 Ideal Case

The most ideal scenario is mathematically described in (5.3), the packet interarrival time is not affected by the virtual router processing and forwarding time.

$$T_k^{in,i} = T_k^{out,i}, \forall i \tag{5.3}$$

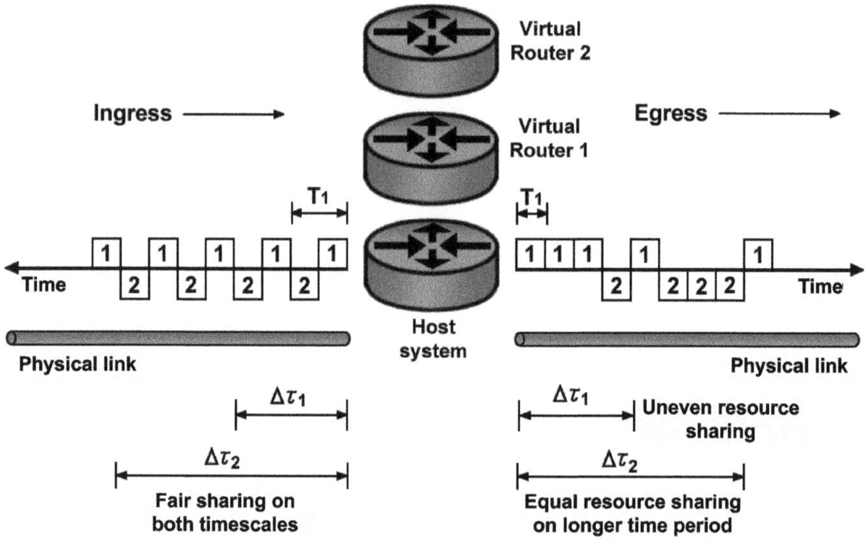


**Figure 5.3** Model and random variables for the multitimescale, comparative disturbance analysis for vitalization in the sharing model.

Without loss of generality, this ideal outcome is achievable when the virtual router processing time and forwarding delay for each packet that flows through the resource is constant (5.4).

$$t_k^{out,i} = t_k^{in,i} + constant \tag{5.4}$$

The condition in (5.4) is, however, unlikely in practice due to a number of factors including indirection, resource sharing, and virtualization that are applied at various system layers when implementing virtual routers. Such a simplistic evaluation is therefore, unfortunately, impractical, requiring other more realistic calculations.

### 5.2.4 A More Realistic Description of the Impact

A realistic evaluation of the impact caused by the virtualization mechanism needs to assess the perturbance with a statistically meaningful number of occurrences (i.e., using a significant number of events such as packets and their interarrival times). It is customary to define an interval of duration $\Delta\tau$ (i.e., the timescale) and to quantify the perturbance caused by the virtual router on the packet arrival/departure times during this period. The timescale $\Delta\tau$ must be chosen to provide a statistically significant number of events and its practical value is discussed shortly. We must first define the statistics used in perturbance analysis.

Figure 5.3 shows various observation intervals of duration $\Delta\tau_{l,k}^{[in|out]}$, and in the top part the abstraction of the event process (i.e., the packet arrival and departure processes). The index $l$ specifies the considered interval. A counter $X_k^{[in|out]}(t)$ is defined as the number of event occurrences during the chosen timescale $\Delta\tau_{l,k}^{[in|out]}$. For the definition of the statistics on the intervals, we start with the determination of the number of packets or events in an interval. We introduce a RV that counts the events (e.g., packet arrivals) for a stream ([*in*|*out*]) up to a certain point of time t:

$$X_k^{[in|out]}(t) = \sum_{t_k^{[in|out],i} \le t;\forall i} \Psi_k^{[in|out]}\left(t_k^{[in|out],i}\right) \tag{5.5}$$

The function $\Psi_k^{[in|out]}(t)$, defined below, selects the packets belonging to a specific stream with index k:

$$\Psi_k^{[in|out]}(t) = \begin{cases} 1, \text{ if event at } t \text{ on stream k during on } [in \mid out] \\ 0, \text{ if no event at } t \text{ on k stream on } [in \mid out] \end{cases}$$

The RV for counting the events $X_k^{[in|out]}(t)$ is depicted in the upper part of Figure 5.3. There are two random variables, one for counting the events at the ingress for flow 1 ($X_1^{in}(t)$) and one for counting the events for the same flow 1 at the egress ($X_1^{out}(t)$).

Subsequently, the counting RV $X_k^{[in|out]}(t)$ can be used to define a random variable $R_{k,j,\Delta\tau}^{[in|out]}$ for the packet arrival and departure rate. This RV is considered for a specific interval $j$ of duration $\Delta\tau$ for an event stream either at the input (i.e., "in," respectively, "ingress") or at the output (i.e., "out," respectively, "egress"). It is

$$R_{k,j,\Delta\tau}^{[in|out]} = \frac{X_k^{[in|out]}(j\Delta\tau) - X_k^{[in|out]}((j-1)\Delta\tau)}{\Delta\tau} \tag{5.6}$$

This random variable on the arrival rate is built on the view that the stream of events is overlayed with a grid of $J$ consecutive intervals of equal duration $\Delta\tau$. The upper part of Figure 5.3 shows a simplified grid. There are two intervals overlayed on the event streams (i.e., packet events) at both the in and out stream.

A remaining question when applying the metric in practice is when to start the intervals in a stream. One can assume without limiting the generality that the first event in the stream triggers the start of the first interval.

This synchronization scheme does not affect the statistics of since intervals are considered ($j$ is very large; $t \rightarrow \infty$) and the event process is memoryless [12].

One can define the average arrival rate of events in intervals of size $\Delta\tau$ for $J$ consecutive intervals:

$$E\left[R_{k,J,\Delta\tau}^{[in|out]}\right] = \left(\sum_{j=1}^{J} R_{k,j,\Delta\tau}^{[in|out]}\right) \Big/ J \tag{5.7}$$

Consequently, the standard deviation of the event arrival rate for $J$ consecutive intervals of duration $\Delta\tau$ is computed as

$$\sigma_{k,R,\Delta\tau}^{[in|out]} = \sqrt{VAR\left[R_{k,J,\Delta\tau}^{[in|out]}\right]} = \sqrt{E\left[(R_{k,J,\Delta\tau}^{[in|out]})^2\right] - \left(E\left[R_{k,J,\Delta\tau}^{[in|out]}\right]\right)^2} \tag{5.8}$$

It should be noted that we consider the mean of a squared RV in the usual way as

$$E\left[X^2\right] = E[X \bullet X] \tag{5.9}$$

Now the *coefficient of variation* results:

$$cov_{k,R,\Delta\tau}^{[in|out]} = \frac{\sigma_{k,R,\Delta\tau}^{[in|out]}}{E\left[R_{k,J,\Delta\tau}^{[in|out]}\right]} \tag{5.10}$$

Note that the denominator of the covariance in (5.10) can theoretically approach zero in steady state. This would lead to an infinite value for the covariance (i.e., it is undefined). However, a shared resource is always occupied for a minimum time. Thus, this minimum prevents the denominator from becoming zero and resulting in the covariance being undefined.

The comparison of the two coefficients of variation for the event rate $R$ for two event streams [in|out] when considering the same intervals $\Delta\tau$ leads to the definition of a *metric for the perturbance introduced by virtualization in sharing mode*:

$$v(R, \Delta\tau) = cov_{k,R,\Delta\tau}^{out} - cov_{k,R,\Delta\tau}^{in} \tag{5.11}$$

### 5.2.5 Smallest Timescale and Timescale Analysis

The timescale $\Delta\tau$ must be set to be larger than the minimal possible interevent time (e.g., larger than the minimal inter-arrival times of packets) in order to avoid pathological mathematical cases (e.g., a packet stretches over multiple intervals). With this requirement satisfied, the timescale can be chosen arbitrarily, at least in principle. Statistically significant results can be obtained if the timescale is above a certain duration (i.e., such that a sufficiently large number of events occur in the considered interval). While equations to guarantee statistical significance are hard to provide, we describe next a practical heuristic to size the timescale.

The lower limit of the timescale should be chosen to the typical rule-of-thumb that five to ten events fall into the interval. Intervals that are too large tend to generate more uniform outcomes due to the law of large numbers. Hence, we suggest this guideline for the scaling of the interval that is based on the expected interarrival time $E\left[T_k^{in,i}\right]$ of events:

$$\Delta\tau \in 2^n \bullet E\left[T_k^{in,i}\right] \tag{5.12}$$

Equation (5.11) describes a scaling behavior proportional the *n*th power of 2. Other scaling behaviors are of course also possible, such as choosing an arbitrary minimal time and scale the interval for the analysis in a linear way. All in all, a timescale analysis should make use of incrementally larger intervals until the law of large number nullifies the effect.

#### 5.2.5.1 Verification of the Perturbance Model

At first, one needs to ask whether such type of perturbances by virtual resources in sharing mode occur in real-world systems. A verification experiment was conducted where both the occurrence and the capabilities of the analysis are verified; the setup is as in Figure 5.2.

The technical details of the scenario are found in [13]. The setup comprises a host server that runs four virtual routers and a single ingress and egress link. The link is shared, and the packets are forwarded from ingress to egress. This combination of the link and the forwarding introduce in general the perturbances on the events as seen at the egress flow and assumed above. To compare the behavior, monitoring points ("duplicators") are introduced that forward the packets to a capturing unit. The capturing unit assigns accurate timestamps for the observations of the packet arrivals. The use of a single capturing unit removes the need for synchronization of distributed clocks. The capturing unit produces log files that can be analyzed offline and are used for calculation of the statistical objects.

A packet generator is used to create four synchronized packet streams. All the streams have packets of the same size, and the synchronization creates an overall stream of back-to-back packets in regular and repeating patterns on the ingress link.

Figure 5.4 shows the observed packet arrivals and the order of the packets on the ingress link during the experiment. A regular grey-scale pattern in Figure 5.4 depicts a well-ordered sequence of packets on the ingress link. Figure 5.5 shows the aggregated flow after leaving the physical router (i.e., on the egress link). The grey-scale order is lost on the egress link. This fact means that the packets on the egress are reordered from the original sequence. One can conclude that reordering and disturbance of the virtual resources is a real-world effect occurring in virtualized systems. In addition, Figure 5.6 shows the throughput for each of the four virtual streams seen at the egress. The variation of the throughput is clearly visible and self-evident. This throughput variation provides the fundamental idea for the subsequent derivation of the performance metric.

### 5.2.6 Capabilities and Conclusion

The capabilities of the suggested concept for performance evaluation of virtualization is demonstrated by Figure 5.7. It shows the change of the metric, such as the change of the metric of (5.11), for example, the coefficient of variation for different timescales ($x$-axis) and when using two virtualization environments (Xen and Virtualbox). Figure 5.7 shows that the change of the metric becomes

| Time epoch | Source IP address | Destination IP address |
|---|---|---|
| 1 | 10.3.1.1 | 10.3.2.1 |
| 2 | 10.0.1.1 | 10.0.2.1 |
| 3 | 10.1.1.1 | 10.1.2.1 |
| 4 | 10.2.1.1 | 10.2.2.1 |
| 5 | 10.3.1.1 | 10.3.2.1 |
| 6 | 10.0.1.1 | 10.0.2.1 |
| 7 | 10.1.1.1 | 10.1.2.1 |
| 8 | 10.2.1.1 | 10.2.2.1 |
| 9 | 10.3.1.1 | 10.3.2.1 |
| 10 | 10.0.1.1 | 10.0.2.1 |
| 11 | 10.1.1.1 | 10.1.2.1 |
| 12 | 10.2.1.1 | 10.2.2.1 |
| 13 | 10.3.1.1 | 10.3.2.1 |
| 14 | 10.0.1.1 | 10.0.2.1 |
| 15 | 10.1.1.1 | 10.1.2.1 |
| 16 | 10.2.1.1 | 10.2.2.1 |

**Figure 5.4** Ingress packet order.

| Time epoch | Source IP address | Destination IP address |
|---|---|---|
| 1 | 10.3.1.1 | 10.3.2.1 |
| 2 | 10.0.1.1 | 10.0.2.1 |
| 3 | 10.0.1.1 | 10.0.2.1 |
| 4 | 10.0.1.1 | 10.0.2.1 |
| 5 | 10.2.1.1 | 10.2.2.1 |
| 6 | 10.1.1.1 | 10.1.2.1 |
| 7 | 10.1.1.1 | 10.1.2.1 |
| 8 | 10.2.1.1 | 10.2.2.1 |
| 9 | 10.2.1.1 | 10.2.2.1 |
| 10 | 10.2.1.1 | 10.2.2.1 |
| 11 | 10.3.1.1 | 10.3.2.1 |
| 12 | 10.3.1.1 | 10.3.2.1 |
| 13 | 10.3.1.1 | 10.3.2.1 |
| 14 | 10.3.1.1 | 10.3.2.1 |
| 15 | 10.1.1.1 | 10.1.2.1 |
| 16 | 10.1.1.1 | 10.1.2.1 |

**Figure 5.5** Egress packet order.

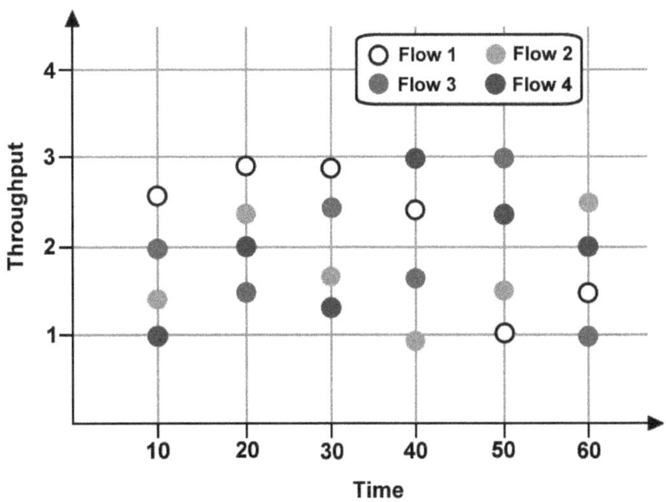


**Figure 5.6** Packet flow throughput ordering.

smaller for longer timescales. This is the expected statistical behavior due to longer time intervals that average out deviations.

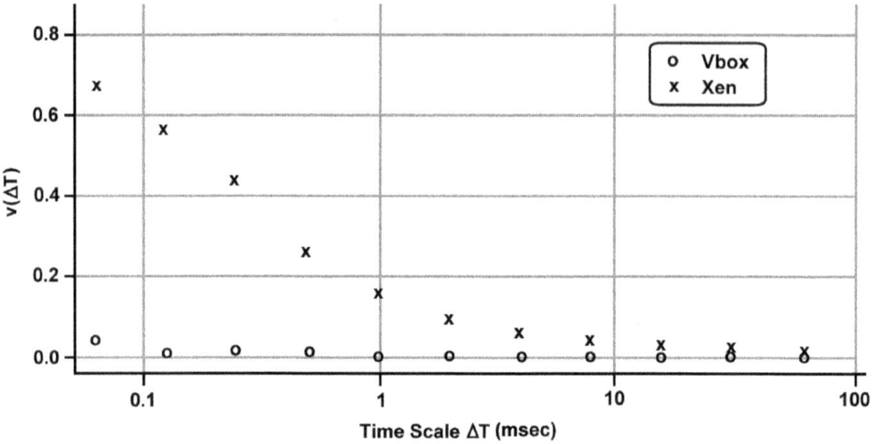


**Figure 5.7** Timescale analysis for virtualization in sharing mode by different virtualization environments.

Figure 5.7 shows the timescale analysis for virtualization in sharing mode by different virtualization environments. The timescale is depicted on the *x*-axis and the metric is shown on the *y*-axis. Figure 5.7 demonstrates that there are significant differences on smaller timescales for the metric (i.e., for the change of coefficient of variation for the different virtualization environments). The differences are almost exponentially larger when using small timescales. This behavior permits the conclusion that certain virtualization frameworks (e.g., Xen in this case) differ significantly in their quality of maintaining packet throughput (i.e., managing the sharing of a bottleneck). The difference among them can be expressed in a numerical way by using the metric in (5.11). Moreover, the metric is independent of the applied mechanisms.

The metric concept we've used here has its own limitation. The analysis does not pinpoint the exact mechanism that causes the disturbance. This knowledge can be obtained only with more insight into the implementation of the sharing mechanism, for example, by applied queuing mechanisms that are used to meet processing due dates and datelines. Such an analysis is probably of high interest for the design of a virtualization mechanisms but is probably also too time consuming or even impossible if the information is not available when designing a complete virtualized system. In its current implementation, the metric provides an opportunity to achieve a rapid numerical evaluation of a virtualization quality while the virtualization architecture is being designed. The metric enables the designer to select the best virtualization method and to look for trade-offs.

## 5.3 Performance Evaluation Concepts for the Aggregation of Resources

Today's CPUs, and consequently, modern servers, scale their performance through massive capabilities for parallelism and by applying many-core computing architectures. This approach is intrinsically an aggregation scheme where resources are bundled together and a single virtual resource is composed of multiple atomics ones (i.e., the available cores).

In this section, we review the performance fundamentals and evaluation concepts for resource aggregation for multicore processors. At first, we detail the foundations using concepts of parallelism (Amdahl's law) and queuing theory. Later we detail two main performance management mechanisms in multicore systems, CPU pinning and NUMA, respectively. Each will be discussed in this order, and since they are tightly related some may find it more convenient to consider NUMA then CPU pinning; we have chosen to address CPU pinning first because in our view the multiple core CPU came into existence first. In either case, both must be understood.

### 5.3.1 Foundations

Bundling atomic resources into a larger system, such as a CPU consisting of multiple cores, is a fundamental concept of modern computer system design. This has been applied since the early days of computer systems and builds on the intuitive understanding that the performance of a system growths when more resources are added (i.e., when parallelism is used). A typical interpretation of the paradigm is that the performance of a system increases by the same amount as the single resource that is added to it. In other words, the performance scales linearly in the number of available atomic resources.

Although this concept sounds intuitive, the actual increase in performance by an added resource is not straightforward and is also difficult to predict. Major conditions for the intuitive view are the assumptions that, first, the resource requests are independent and don't influence each other, and second, that the amount of computation needed to manage the requests (e.g., the work for the assignment to a resource) is negligible.

Several computational problems, particularly the class of parallel ones, can be split in multiple, independent, and smaller computational requests. However, many computing problems share similar resources (e.g., cache, memory, I/O, busses, and data) and therefore violate the independence requirement.

Next, we describe and review two well-established and fundamental methods, Amdahl law's and processor sharing queuing models, for the performance prediction and evaluation of computer systems that allows parallelism.

#### 5.3.1.1 Amdahl's Law

Gene Amdahl was a pioneer in early computer system engineering and a key person in the design of IBM's /360 mainframe system (see Chapter 3). Amdahl formulated his law and insights in computer system performance in 1967. Since then, his observation has become an important guideline for resource engineering in computer systems at large.

Amdahl's law describes the theoretical improvement of system performance when a certain fixed workload is served by parallel resources [14]. The law's main idea is that when one speeds up one part of a system, the effect on the workload performance depends on both how significant this part was and how much it sped up [15]. Therefore, we consider a system in which executing a certain workload requires time $T_{old}$. Suppose now that a fraction $\alpha$ of this load can be accelerated by parallelism on multiple cores and its performance is improved by a factor of k (i.e., using *k* core). The time that the workload originally required was $\alpha \bullet T_{old}$ and now it requires time $(\alpha \bullet T_{old})/k$. Hence, the new overall execution time $T_{new}$ calculates to

$$T_{new} = (1-\alpha)\bullet T_{old} + \frac{\alpha \bullet T_{old}}{k} = T_{old}\left[(1-\alpha)+\alpha / k\right] \tag{5.13}$$

This equation allows the computation of the speedup $S = T_{old}/T_{new}$ as

$$S = \frac{1}{(1-\alpha)+\alpha / k} \tag{5.14}$$

*Example:* We provide a brief numerical example for a better understanding of the law. We assume that the part of the task that is parallelized is initially 75% of the time $(\alpha \bullet T_{old})$ and has an acceleration factor of $k = 4$. This results in a speedup of $1/\left[0.25+\frac{0.75}{4}\right] = 2.29$. Although the system now uses four parallel computing resources and a rather large part of the task can be accelerated, the overall speedup was significantly lower than the parallelism may imply.

The awareness that just adding parallelism and resources to a system, for example when bundling resource into a virtual one, does not automatically increase the system performance by the same amount is a major message of Amdahl's law. Figure 5.8 shows the asymptotic behavior of Amdahl's law for very large numbers of parallelism (i.e., processors or cores), and various portions of the task that can be parallelized.

One other way of viewing this is a quote from Ken Batcher [16]: "parallel processing is a method of converting a compute bound problem in to an I/O bound problem."

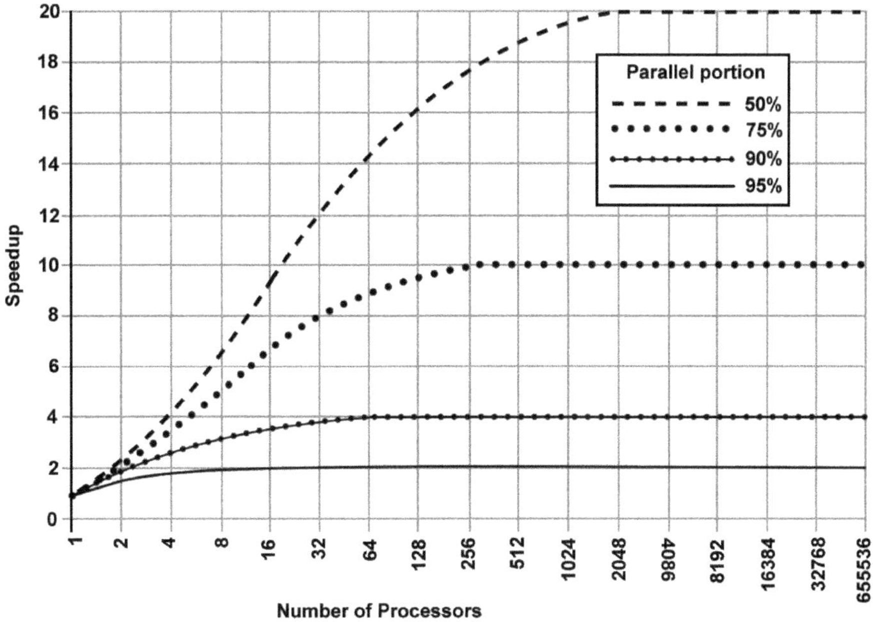


**Figure 5.8** Amdahl's law: theoretical speedup of the execution of a task as a function of the number of parallel computing elements.

### 5.3.1.2 Queuing Theory Models

The bundling of server resources for achieving performance is one of the core themes in queuing theory. The concept can analyze the performance of an array of parallel servers for a stream of requests. Basically, the modeling by queuing theory might assume that either K servers are available for a single queue or that K jobs simultaneously share a single resource or system. Here, we summarize selected results for the so-called M/G/1/K*PS queue, which are queues where K jobs are in the system and share a processor. We base our description on the work in [17].

The model for the considered M/G/1/K*PS is depicted in Figure 5.9. The requests arrive according to a Poisson process with rate $\lambda$. The average service requirement of each request is $\tilde{x}$. The service can handle at most $K$ requests at a time. A request will be blocked if the number has been reached. The probability of blocking is denoted as $P_b$. Therefore, the rate of blocked requests is given by $\lambda P_b$.

The probability mass function of the total number of jobs in the system has the following expression [18] with $\rho$ as the offered traffic by the stream:

$$P[N=n]=\frac{(1-\rho)\,p^{n}}{\left(1-p^{K+1}\right)} \tag{5.15}$$

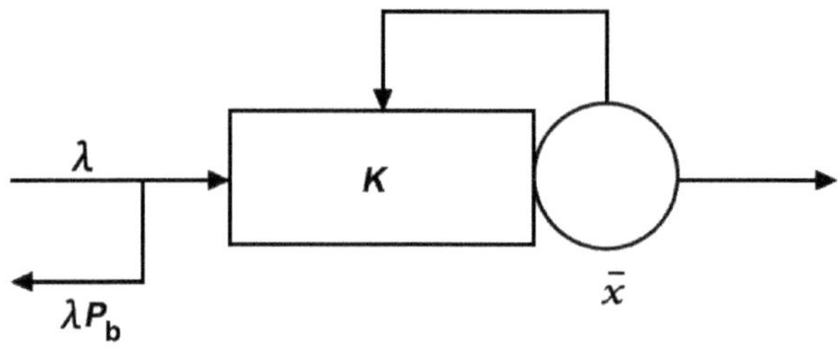


**Figure 5.9** An M/G/1/K*PS queuing (e.g., for web server).

From (5.15), we can derive the following three performance metrics: the average response time $T$, the throughput $H$, and the blocking probability $P_b$ as

$$P_b = P[N = K] = \frac{(1-\rho)\,p^{K}}{\left(1-p^{K+1}\right)} \tag{5.16}$$

$$H = \lambda(1 - P_b) \tag{5.17}$$

$$T = \frac{E[N]}{H} = \frac{\rho^{K+1}(K\rho - K - 1) + \rho}{\lambda\left(1-\rho^{k}\right)(1-\rho)}$$

Using these results and estimating the parameters for $\tilde{x}$ and $K$, one can investigate the system behavior when increasing the arrival rate $\lambda$ for requests. The computed average response times are shown in Figure 5.10, throughput in Figure 5.11, and blocking probability in Figure 5.12. The blocking probability in Figure 5.12 shows the asymptotic behavior of the system when increasing the arrival rate. In addition, part (a) of the figure outlines the limiting of the average response time and of obtainable throughput.

## 5.4 CPU Pinning

CPU pinning may be more accurately described as core pinning within a CPU. CPUs today, specifically those found in server-class machines, will have multiple cores per CPU. In some cases, the number of cores today is reaching several scores. Modern AMD CPUs are rapidly approaching the triple-digit values for the number of cores, and the competing Intel CPUs are not far behind. While we often focus on, or at least lead, with a discussion on the number of cores in modern CPU, by some estimates the modern server CPU may have as much as

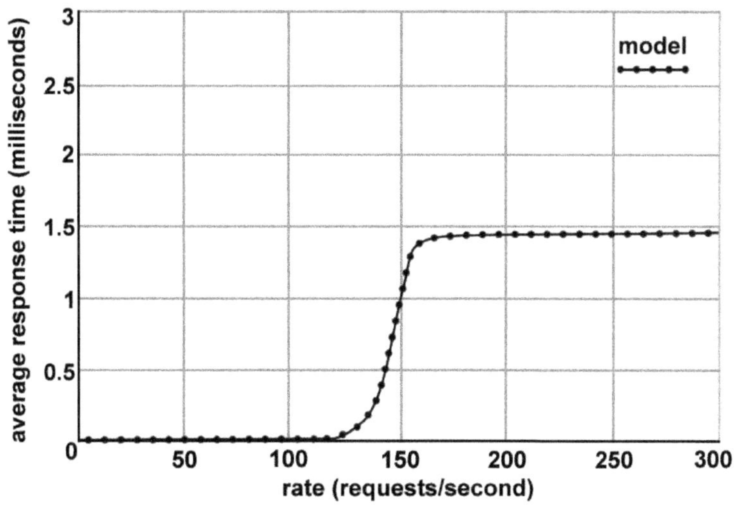


**Figure 5.10** Compound average response time.

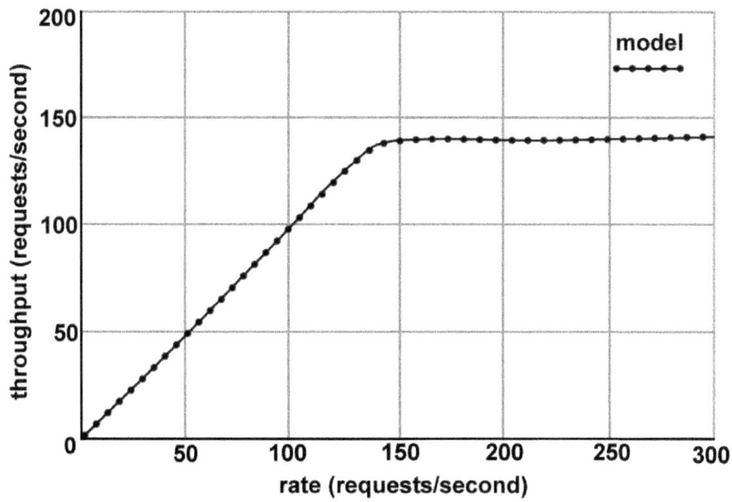


**Figure 5.11** Compound average throughput.

40% of the total silicon die area dedicated to functions other than the compute functions found in a core. It is the relationship between these cores and the remainder of the functions on the CPU that, depending on the functions being performed, result in the benefits from CPU pinning.

We are talking about the heavy metal CPUs found in servers. These are significantly different than their smaller counterparts that may be found in client machines where we may find only one or two handfuls of cores. There may

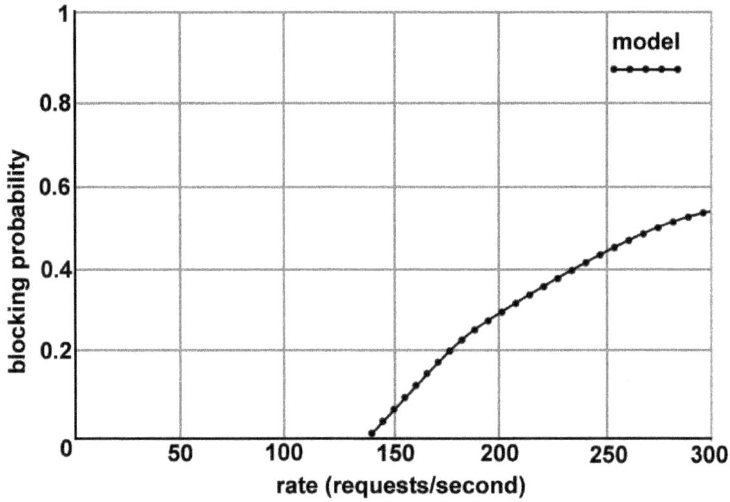


**Figure 5.12** Compound blocking probability.

be other differences in the actual cores also, for example the server class CPUs may support more complex instructions natively that are not present in the client CPU core.

The CPU as a minimum will have Layer 3 and last layer cache, memory management and address translation and I/O functions built into the die or dies.

Depending on the final architecture, the last layer cache may be shared across multiple cores or dedicated to a single core. If the last layer cache is shared, then more than one core will be capable of addressing the same cache memory space. Fundamentally there are two types of information: involatile instructions and data, which can be changed. Both need to be close to the core in order for the information to be acted on by the core. The old adage is that information wants to be close to the core. If this last layer cache is shared between multiple cores, then there is a need to serialize the access to prevent the two cores from causing some type of conflict; otherwise, the address being generated or the data being written can cause an undesired outcome. If the running program on the core generates a request (or is likely to generate a request) to an address space for any information (instruction or data) and that information is not currently contained in the last layer cache, then the memory management portions of the CPU are engaged, the core is possibly stalled (pauses any processing), and the cache is updated with the required information. Further back into the memory hierarchy is a shared Layer 3 cache, which is typically a much larger cache than the last layer cache yet significantly smaller than the total populated CPU memory. Again, if the Layer 3 cache does not contain the required information the memory management on the CPU will update a page or pages (memory is segmented into smaller blocks called pages).

Modern operating systems, the most common being an instance that relies on the Linux kernel and a large collection of supporting libraries and applications, have long supported processes (a running instance of a program) with multiple threads. These operating systems are capable of supporting multiple lightweight processes. These lightweight processes are often called "threads," likely due to some of the early library API calls allowing for their creation. In addition, the silicon in the core on the CPU may also support multiple hardware threads. One point of caution; the term "thread" can be overloaded. A single core of a CPU may be capable of supporting one or several hardware threads that are within scope at any given time. One or multiple processes in scope will consist of at least one software thread and possibly a very large number of software threads. Here in this example, we have intentionally overloaded hardware threads with software or application threads to demonstrate our point. A hardware thread may be distinguished from a software thread by discussing the latter as a lightweight process for clarity in some contexts.

The concept of CPU pinning locks, via some mechanism native to the kernel, an instance of a running lightwight process to a specific CPU core; this prevents the core from being used by any other lightweight process for the duration of the pinning (or life of the lightweight process), and also prevents the operating system from scheduling any other task to that core. This provides several advantages in virtualized applications; one is that the core and the associated last layer cache are significantly resistant to suffering a delay due to context switching of the core to other tasks, and another is that if selected correctly, the core can be aligned to other resources on the CPU, such as I/O (e.g., network cards), as will be discussed in the next section.

## 5.5 Non-Uniform Memory Access

NUMA can be found in both single-socket and multiple-socket servers. In single-socket servers NUMA is found within the CPU itself, and in any multiple-CPU server may be found in the CPU and the server design.

The two primary considerations when considering NUMA are the memory architecture of the CPU, and by extension of the server, and the I/O (mostly the network interface card) configuration.

Today many servers are multi-CPU, and in the telco space these are typically dual-socket servers; that is, there are two CPUs in each server. Some single-socket servers are emerging and may still have NUMA considerations due to a chiplet design within the CPU.

Various server vendors will design the I/O either in a symmetrical or unsymmetrical configuration and also may populate the interface cards from the factory in certain ways that have NUMA implications. Nearly all known standard high-volume servers available today use the PCIe bus to accommodate the NICs. These PCIe busses (there are a multitude) are directly associated with specific sockets on the main board and are thusly associated singularly to one of the CPUs. Similarly, the total system memory is not all on a single bus, but rather allocated in banks to specific CPUs. Typically, the vendor will require (but not always mandate) that all the system memory be installed in a balanced configuration. For example, on a system with 128 GB of main memory in a two-socket server, each CPU will have direct access to 64 GB of main memory and the full memory access of the server by any one core will not be direct.

The main board for the I/O systems on the other hand may be designed (it would be our view that this is a less than optimal design decision) unsymmetrically. For example, it is not uncommon for systems to be designed to support disk I/O on one CPU and network I/O cards on the alternate CPU in a dual-socket system. In fact, in many pure cloud server architectures this can be argued as an advantage, but in latency and jitter constrained systems, not so much.

When discussing NUMA alignment, the concern is often associated with workloads with continuous and heavy network traffic. This implies that the NIC card (and the buses on which it resides) are plumbed to the same CPU as the software workload. Thus, to take full advantage of both CPUs in a multiple CPU system the system would, out of necessity, have dual NIC cards. Each NIC card plumbed to the PCIe is associated with each CPU even though these CPUs are interconnected on the server by and internal CPU to CPU bus, such as the Intel UPI™ bus. Otherwise, the traffic from NIC to CPU has to pass over the internal CPU bus, producing less than optimal latency and jitter predictability.

CSP workloads by nature are not highly dependent on the disk I/O and therefore there is not a need to expect or demand NUMA alignment for purposes of disk I/O.

While the NUMA alignment of memory is also one area of possible concern, both the typical balanced nature of the memory due to vendor requirements, and given the tendency of modern operating systems to be memory-aware, the likelihood of having unpredictable performance due to poor main memory alignment are mitigated in most cases.

This may not be the case though when looking at the cache memory considerations within the CPU itself, and there are also NUMA considerations here.

Consider Figure 5.13 as a server where the I/O for PCIe cards is balanced to each socket, and Figures 5.14 and 5.15 represent two different CPU architectures with different NUMA considerations.

We actually have at least three layers of memory: main memory, L3, and last layer (sometimes called L2 cache). The core can only access data and instructions (called the object for simplicity) from the last layer cache. If the object being retrieved is not in the L2 cache, then the L3 cache must update the L2 cache, and in some architectures this can happen simultaneously. If the object does not reside in L3, then a page must be retrieved from the main memory. This action will stall the processing within the core until the L2 cache has the desired targeted object. There are differences in the two approaches shown in Figures 5.14 and 5.15. In the architecture of Figure 5.14, internal tests have shown that a core with dedicated L2 and L3 caches provides predictable latency and jitter when a cache miss is encountered. In contrast, the architecture in Figure 5.15 results in a jitter pattern that is uncontrollable with other tasks running on other cores and contention with the memory cross connect shown in Figure 5.15.

This discussion may serve as stimulus for research on the internals of core pinning and NUMA alignment as these details are still being investigated privately by vendors of servers and CPUs.

One proposed solution, which may address part of but by no means the entirety of this issue, is to simply use a single-socket server. In fact, some deployments have been witnessed where servers designed to accommodate dual-socketed CPUs have been deployed with only a single socket populated, provided all the I/O can be plumbed to the sole socketed CPU. There are some concerns here too of course, one being that the CPU purchased has die area, however

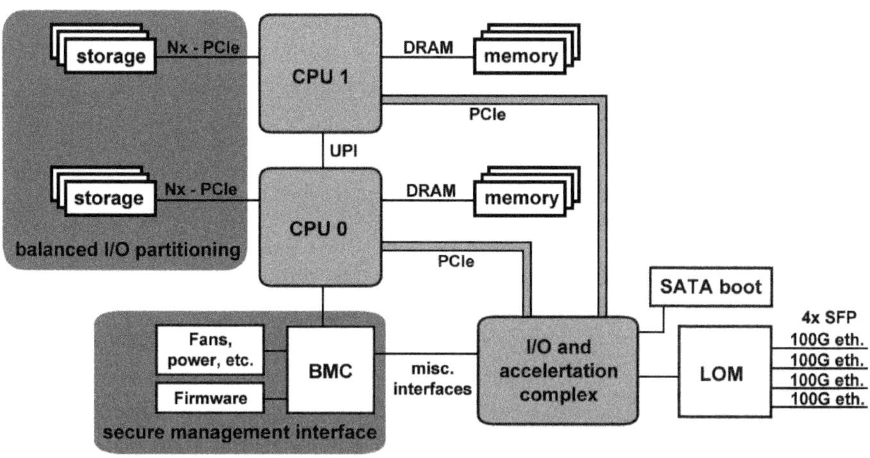


**Figure 5.13** A NUMA aligned dual-socket server. Note the balanced PCIe on each CPU.

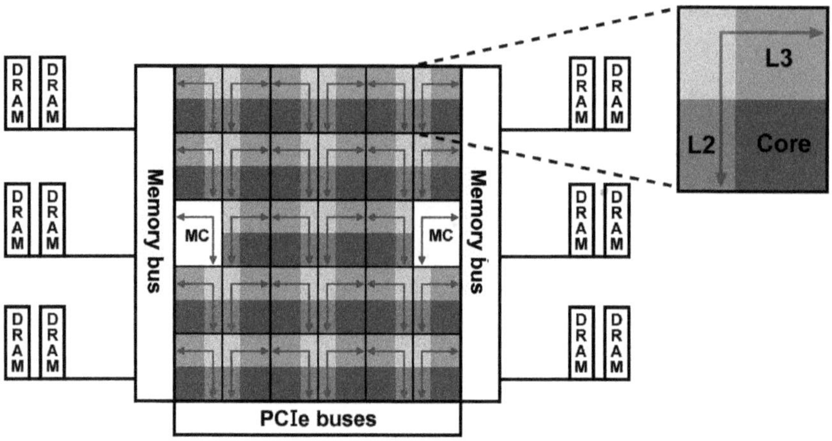


**Figure 5.14** NUMA aligned memory intra-CPU.

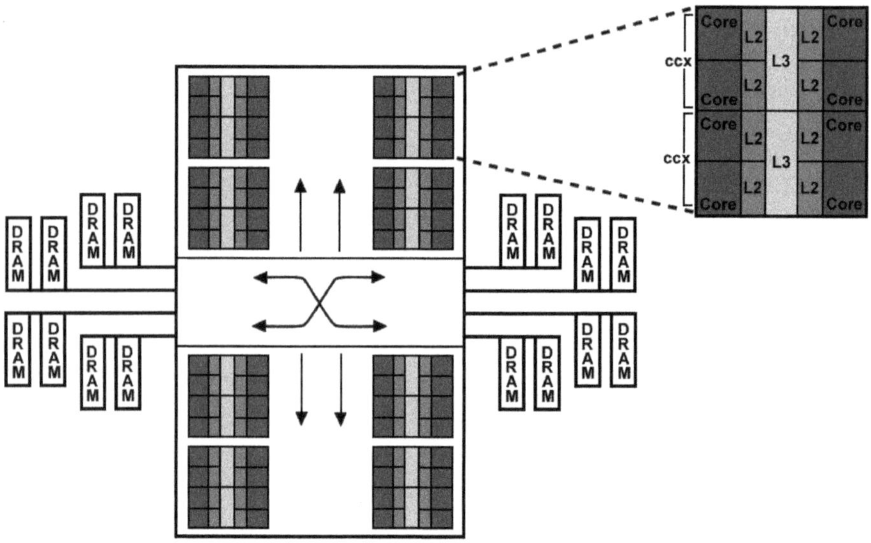


**Figure 5.15** Non-NUMA aligned memory intra-CPU.

small, allocated to the CPU-CPU interface that is now wasted. Another concern is that the server consumes more real estate in a rack than necessary for the performance that will be delivered by the server. One final concern is that the abstract concept of using standard high-volume servers may be weakened. Typical cloud consumption volumes of these servers significantly outnumber the consumption of the same servers to be used in a dedicated CSP environment. A question of scale (e.g., the economy of scale) may be lost if the CSP and the cloud or hyperscaler do not consume similar systems.

## 5.6 Conclusion

Today's standard high-volume servers used in CSP virtualized deployments are very complex systems from the CPU design to I/O configuration, and the software running on them in a virtualized environment brings additional complexities. A well-designed solution requires that the designers of the software have a detailed understanding of the CSP networking requirements and the underlying design of the modern CPU and the servers in order to fully optimize the solution for CSP deployment. This insight by the operators is part of their effort to transform the way they operate their network using individuals that have software, system design, and hybrid networking skills. This is an ongoing effort by many of the operators to transform their operations staff. While running the code and taking some measurements may satisfy the functional requirements, ignoring the details of the server and CPU may lead to product performance or cost issues that other competitors can exploit to their advantage.

## References

[1] Lampson, B., "Hints and Principles for Computer System Design," https://www.microsoft.com/en-us/research/uploads/ prod/2019/09/Hints-and-Principles-v1-full.pdf.

[2] Russinovich, M., M. Costa, C. Fournet, et al., "Toward Confidential Cloud Computing," *Communications of the ACM,* Vol. 64, No. 6, 2021.

[3] DPDK Project, https://www.dpdk.org.

[4] VMWare Workstation; https://www.vmware.com/products/workstation-pro.html.

[5] Oracle VirtualBox, https://www.virtualbox.org.

[6] iptables, http://git.netfilter.org/iptables/.

[7] Kohler, E., R. Morris, B. Chen, J. Jannotti, and M. F. Kaashoe, "The Click Modular Router," *ACM Transactions on Computer Systems (TOCS),* Vol. 18, No. 3, 2000.

[8] Fiedler, M., K. Tutschku, P. Carlsson, and A. Nilsson, "Identification of Performance Degradation in IP networks Using Throughput Statistics," *Proceedings of the 18th International Teletraffic Congress-ITC-18,* Berlin, Elsevier, Volume 5, 2003.

[9] Stezenbach, D., and K. Tutschku, "A Performance Evaluation Metric for NFV Elements on Multiple Timescales," in *IEEE Global Communications Conference (GLOBECOM),* Atlanta, GA, 2013, pp. 516–521.

[10] Fiedler, M., T. Hossfeld, and P. Tran-Gia, "A Generic Quantitative Relationship between Quality of Experience and Quality of Service," *IEEE Network,* Vol. 24, No. 2, March-April 2010, pp. 36–41.

[11] Matragi, W., K. Sohraby, and C. Bisdikian, "Jitter Calculus in ATM Networks: Multiple Nodes," *IEEE/ACM Transactions on Networking,* Vol. 5, No. 1, February 1997, pp. 122–133.

[12] Bryant, R., and D. O'Hallaron, *Computer Systems: A Programmer's Perspective,* Third Edition, Boston: Pearson Education, 2016, pp. 22–26.

[13] Feller, W., *An Introduction to Probability Theory and Its Applications,* Volumes 1 and 2, John Wiley & Sons, 2008.

[14] Kleinrock, L., *Queueing Systems,* Volumes I and II, New York: John Wiley & Sons, 1975.

[15] Harchol-Balter, M., *Performance Modeling and Design of computer Systems: Queueing Theory in Action,* New York: Cambridge University Press, 2013.

[16] https://en.wikipedia.org/wiki/Ken_Batcher.

[17] Cao, J., et al., "Web Server Performance Modeling Using an M/G/1/K* PS Queue," in *10th International Conference on Telecommunications (ICT 2003),* Vol. 2, IEEE, 2003.

# Part II
# Engineering of Virtualized 5G and B5G Systems

# 6

# Transforming and Disaggregation in 5G and B5G Networks

## 6.1 The Transforming and Disaggregation of the Network

Many mobile operators today have significant investments in 4G network infrastructure (in addition to previous generations of the mobile network) and the introduction of 5G, in many cases, will coexist in the network with existing 4G networks for the foreseeable future. In some cases, even the nodes that are deployed will support both 4G and 5G functionality in the same system. Transforming these systems to a pure virtualized network will take considerable effort from the NEPs in how these systems are developed and how the telcos will deploy and operate these systems. This chapter will provide discussion and analysis of this challenge and the current state of the art, including areas where additional considerations need to be addressed and resolved.

Much of this material is based on existing practice in the leading CSPs production deployments; as such, the research sources from academia are limited, and much of the information presented here is based on either domain knowledge from the field or private conversations. Therefore, readers are encouraged to explore internet search results and possibly acquire private industry studies (e.g., paid reports from analysts).

### 6.1.1 Challenges to Transforming the Telco Network

The mobile network is often separated into two subnetworks, which are the RAN and the core network; Figure 6.1 shows the common separation [1]. There are several reasons that this separation has evolved over time. One reason is the operators will often have separate teams of support staff responsible for the management of these systems in part due to geographic separation or different skill sets. Another reason is that in some very large networks, the RAN may be provided by one vendor, and the core, or elements of the core, may be provided by a different vendor.

In addition, in some very large networks the RAN may be supplied by one vendor in multiple regions and in other regions the RAN may be supplied by a second vendor. Even in the case where the RAN and core are delivered by the same vendor, it is very common for different operational teams at the telco to be responsible for the deployed platforms. In part, this is due to the historical evolution of the mobile network and that the required operational skills to

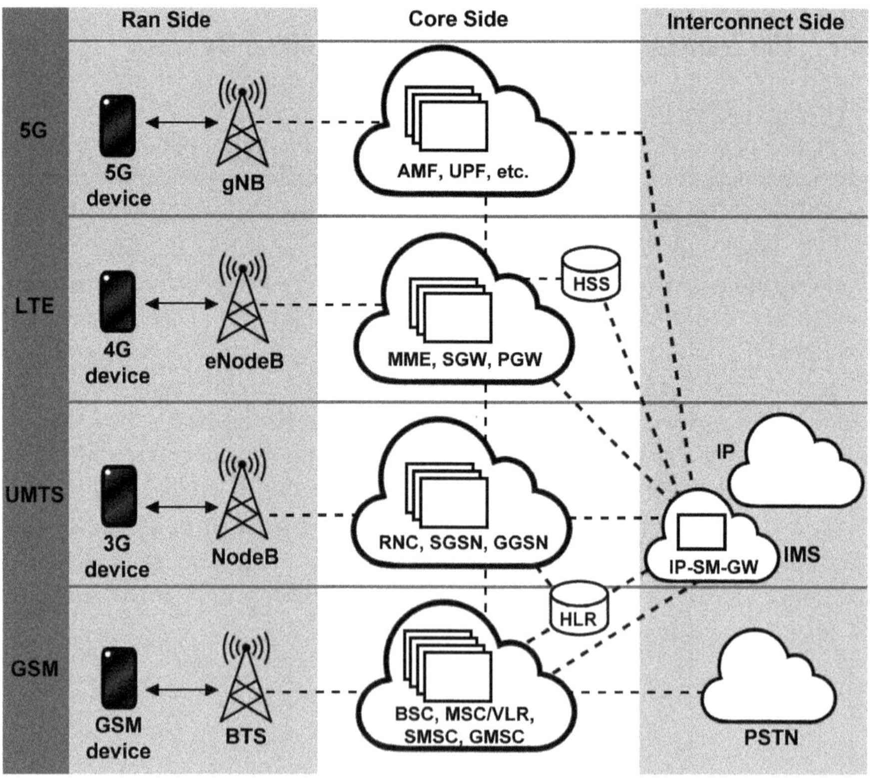


**Figure 6.1** RAN, core, and interconnect sides of the CSP network.

maintain the systems may be different. In addition, and due to the design of legacy purpose-built systems, each vendor may have designed and implemented their system and platform based on different requirements from the engineering teams. For example, there are cases where systems developed on UNIX-derived hosts have had native access to the host via standard tools such as telnet and SSH ports of 23 and 22 intentionally disabled by the vendors to prevent operators from accessing the host operating system natively, even via a management console on the host. While this is rarer today than in the past, it serves to illustrate that vendors may choose from time to time to tie the underlying operating system into their proprietary tools for managing and operating the systems.

## 6.2 DevOps: A Method to Improve System Management

As described earlier, DevOps is a long-established methodology in the cloud and with hyperscalers [2]. In practice the term is also used to describe a culture or mentality combined with practices and procedures for developing and maintaining systems. The intent of DevOps is to merge what had previously been two silos found in traditional IT organizations. These two silos are the organization writing the code that implements a service functionality, and the organization providing day-to-day support of the service once it is deployed. Consider an application using a database running on a cluster of Linux servers; one team would build the application using the database features, and once the functional and performance requirements are met (or nearly met) the development team would release their product to a production team. The production team would then deploy the application on the infrastructure and become responsible for the application in production. At this point in time, it is not unusual for the development team to be moved on to other tasks. This application may be one used either for internal or external customers. Nevertheless, from time to time events might arise that cause the system (the application and all underlying software and hardware) to fail in some way. The operations team would be the first line to triage the event or events, and when possible, use their skills and resources to resolve the event. On occasion, the event may exceed the resources or ability of the operations team to resolve; in this case, they would have to interrupt the current tasks of the development team to address the fault. This interruption could possibly impact developer delivery schedules, or in some extreme cases, cause a manager to have to wake a development engineer in the middle of the night or interrupt a holiday weekend. This model has been in place for decades and the developers and operations teams are aware of the shortcomings. One final example concerns the operating system itself. From time to time security concerns arise where an element of the operating system requires updating to close a security vulnerability. In this case, the operations team may or may not consult with the development team on the appropriate actions necessary to update the operating system to fix the defect. This example will be addressed again later when discussing the telco model of DevOps and the implications for telco operations in a virtualized environment.

The theory of DevOps is straightforward—the team that is building software is also the team responsible for the operations and maintenance of the software and systems on which the production software runs. However, the model has been oversimplified here for brevity. Prior to the adoption of DevOps, even within companies that might be classified as ISVs, there were often two teams of developers. One team was working on the next major release of software, and the other team was supporting the software already in production, on the company's own systems, or with customers who had licensed the software, or in their network. This model developed over many years of practice, and some saw it as creating a two-tiered system within the developer community: one team of developers creating new features and products, and a second team of developers fixing defects left in the wake of the first team. The DevOps model, where this bifurcation of the development teams is eliminated, was introduced into practice by some of the hyperscalers at about the same time as the virtualization revolution. By some accounts, the hyperscalers using the DevOps model were able to accomplish two significant goals: the first was to quickly fix defects, and the second was the ability to introduce new features in production more rapidly. These defect removals and new features would be tested in small enclaves before being rolled out en masse. This also provided an opportunity to significantly reduce the number of people responsible for maintaining the operational network and imposed accountability for efficient and reliable operations on the team developing the software in the first place. By some accounts the hyperscalers are several orders of magnitude more labor-efficient at maintaining their operational systems than the average telco. In addition, whereas a telco may historically have been able to upgrade software in their operational network several times a year, the hyperscalers are able to continuously apply upgrades to their production network. Figure 6.2 shows a classical CSP operations team where there is specific vendor alignment for every product on the left of the figure, and a separated NFVi support team on the right, and only the software is supported by a vendor-specific team. The operational technology (OT) network is where the money is made in the network.

Considering that a significant portion of the TCO of the telco network is related to salary, benefits, and overhead associated with the operations teams, even minor improvements in the operations efficiency can have significant cost benefits.

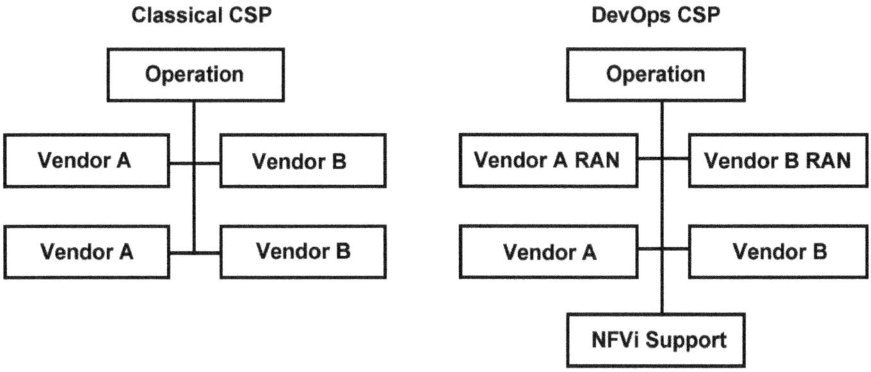


**Figure 6.2** Common CSP OT operational structure.

## 6.3 Telco DevOps

DevOps is a great model for hyperscalers and anyone who develops software for their operational network. It stands in stark contrast to the model used by many telcos that is based on procuring systems and solutions from a number of vendors and then operating the systems in the production network themselves. However, there are significant differences in the operational and customer environments of the hyperscalers and the telcos. Some of these differences are driven by regulatory requirements and critical infrastructure concerns within the telco environment. As a result, hyperscaler DevOps and telco DevOps are different in some key respects.

In the days of purpose-built systems, these platforms consisted of computing elements and an operating system but may have been wrapped by the vendor to discourage the telco from accessing the underlying machine directly. This was not without good reason historically (e.g., preventing misconfigurations or protecting the vendor's intellectual property) but may have outlived its practical usefulness as networks transition to software. One point to consider is that today, the virtualized systems, possibly with some qualified exceptions, are built on some form of Linux. Linux of course has its roots in Unix. One quote from the Official Linux kernel README file even claims that Linux is a Unix clone. This relationship between Linux and Unix is mentioned here simply to provide the historical connection between the two and to base our ongoing discussion on this foundation. Unix was developed at the original AT&T Bell labs back in the late 1960s and found its way (in a number of different forms) into the telco network long before the creation of Linux in 1991. Today, Linux can also be found in a number of different forms and on a number of different systems in the telco network. One may argue several points, including that Linux is royalty-free whereas Unix is not; as well, Linux was initially developed to run on the x86 architecture and Unix was initially designed for different CPU platforms. The point, however, that should be considered is that a system administrator who has a strong foundation in Unix should have little problem transferring their skills to Linux-hosted systems. While there are cases where someone may need be specific about the exact release of the operating systems (usually the kernel) or other middleware on a system, the basic knowledge is transferable between UNIX and Linux.

On the other hand, a system administrator, who was trained using interfaces (command line or graphical user interface (GUI)) where a vendor has created proprietary wrappers around the native UNIX shells and commands, may find that their skills are not as easily transferrable. For example, even legacy Cisco switches and routers had an underlying instance of a heavily modified UNIX operating system and offered a CLI that buffered the user from the operating system. Vendors had very powerful reasons for this buffering; they were responsible for the entire platform that would be in the hands of an operational team not of their choosing. While the vendors could provide extensive training to the systems administrators at an operator, there was nothing to prevent someone from issuing a powerful command that could take down the system or even totally erase the contents of the operational file system. To prevent the possibility of an errant command being issued (the UNIX and Linux command line can be a very powerful capability), the vendors in some cases would remove portions of the operating system and replace it with an interface that protected both the administrator and the vendor. This eventually led to the situation where network functions in the network from one vendor would have a completely different operational interface than the same function from a different vendor.

Some of the largest telcos have insisted on a multivendor approach when deploying their operational network, which is known as not putting all your eggs in one basket. This leads to the case for the example above where the same function in the network from different vendors might require two different support teams with different skills tuned to a specific vendor's implementation and tool set. This can lead to the expansion of their support teams, thus causing an increase in their OpEx.

There are two other operating systems that may be encountered that are not considered to be in the Linux family; these are VMWare and Wind River. While some may argue that they too have their foundation in UNIX, others may argue that they have no relationship to UNIX. Both VMWare and Wind River provide similar capabilities to those found in Linux that are necessary for virtualization of the CSP network elements and both have found their way into production networks. Hence, we will treat them correctly or incorrectly as variants of the original UNIX and fully capable of meeting or exceeding the needs of the CSP production workloads in a virtualized environment. The operational skills may be more difficult to claim as being 1:1 with Linux.

Let's use a very simple analogy here to illustrate this point. Consider two different automobile companies where each has designed a different user interface. The first interface is the one we are familiar with today, and the second interface has controls that may be found in a helicopter. An administrator operating one function in the network would have to be proficient in both interfaces to manage the platforms from the two vendors. In addition, it could require different tools and skills during any maintenance operation. Not an ideal situation. As a result of this concern the telcos developed different teams to manage these platforms from different vendors, even though the core functionality was identical. Furthermore, the actual maintenance and tools used on the systems might be significantly different. One of the promises of NFV and SDN is that while the upper layer software may be from different vendors, and thus require different skills and possibly different staff, the majority of the underlying system (i.e., the network, compute, and storage of the physical system and the NFVi layer (the middle software layers)) are common. Using what today is called a cloud-native model, the telco could reuse the same support teams for any solution in the network up to, but possibly excluding, the actual application layer software [3].

This model should not be a surprise. Commonality of parts and their familiarity was the spark that lit the first industrial revolution. This principle is now being applied to significantly more complex solutions, including software and systems, rather than the wheels of a carriage.

## 6.4 Transforming the Operations in the Network

Transformation considerations can be divided into two segments: first is the physical transformation of the purpose-built systems of the past generations of the mobile network, and second is the organizational transformation of the operators of the network. One is physical based on the design of the equipment (hardware and software), and the other is the transformation of the business processes for building and maintaining the network. The network equipment in 5G and B5G is often separated into two architectural divisions: the radio access network, often simply called the RAN, and the core. Likewise, the operational side of the telco business can also be divided into two categories: existing operators with legacy networks, and new entrants starting with a clean slate. The latter are called green-field operators by some in the industry, as mentioned in Chapter 1. Green-field entrants have no legacy network to maintain and support; there are a very small number of these operators to be found today and some have been mentioned in earlier chapters. Reliance Jio and Rakuten are two large green-field operators that have emerged late in the 4G evolution and early in the 5G timeframe. Dish may also be placed in this group, although some may argue that they are not a green-field operator but rather a transplant coming from the satellite space. This distinction between green and brown field is significant because the existing legacy infrastructure and the organizational structures supporting each significantly influences the risk analysis trade-off and financial cost model for building and operating a new network.

The legacy operators will have long established policies and procedures in place, many that may in fact predate the technology that is being maintained. These policies and procedures have been developed and evolved over a very long time and were based on preventing and quickly resolving events that would adversely impact the operation of the network. As an example, there may be a unique method of procedure (MOP) for any configuration change to every unique element in the network. This ensures that each element in the network has a specific and traceable change history but it adds significantly to the operational overhead of maintaining the network. It may be easy to ask the operator to simply change this procedure, but in practice this increases risk for those individuals responsible for day-to-day operations who are required to deliver service continuity according to a set of defined key performance indicators or customer SLAs. This may be one reason that some hyperscalers have had success lately moving into the space of operating the network. One example was the move by AT&T to transition the management of their 4G network and the associated staff to Microsoft [4]. Moving the staff and responsibility for managing the core to an enterprise outside the traditional CSP and to the hyperscaler brings the possibility that legacy policies and procedures can be challenged and potentially changed to reflect a more cloud-native approach to managing these networks. Some debate remains on the success of this strategy and not all in the industry agree here that success can be claimed. One risk does remain, however; that of migrating the actual core to a hyperscaler entirely. There have been examples of large-scale outages at the hyperscalers that would consume as much as 100 years of downtime when compared to the five nines, also known as 5'9s or 99.999% uptime expected for a telecommunications network. Five nines is equivalent to less than 5 minutes down time per year (this includes scheduled and unscheduled time). A 10-hour outage is the same impact to the uptime as would be allocated to nearly the entire life span of the operational network (a 10-hour outage would be over 120 years' worth of downtime in a five nines requirement). On the other hand, engineering network infrastructures in a software environment opens up new architectural options to deliver fine nines resilience but these are not necessarily fully understood given the maturity of the topic [5].

One final comment on the risk of managing the network: there is a responsibility in the deployment jurisdictions for resilient operation of the portion of the network considered to be critical infrastructure. Outages impacting critical infrastructure can have a financial punitive penalty to the operator and result in regulatory oversight increases. None of this is currently borne by hyperscalers when they sustain a service impacting outage. However, this is likely to change and the first hint is the increasing attention of regulators to hyperscaler failures to comply with local laws relating to data sovereignty, privacy, and speech. As a result, there is institutional resistance within many operators that must be overcome before they fully embrace the cloud-native model of a fully disaggregated and virtual network.

## 6.5 Rolling out 5G in the Network

The introduction of 5G can be broken down into two architectures: non-standalone and standalone. Non-standalone, sometimes called NSA, is applicable to network operators who are also operating a 4G network. This approach allows an existing operator with a 4G core to introduce a 5G RAN that will interoperate with the 4G network core. Release 15 of the 3GPP specifications provided for this capability. Standalone or SA, requires both a 5G RAN and 5G core [6].

### 6.5.1 5G Non-Standalone and Standalone Considerations

There was discussion about the separation of the control plane from the data plane in Chapter 2. The NSA strategy to introduce the 5G New Radio relies on this separation as a starting point and places additional requirements on some of the legacy 4G nodes. 5G-capable user equipment (UE) and 5G New Radio become the two additional essential elements in the NSA model, while the remainder of the core is based on 4G LTE Evolved Packet Core (EPC) standards. This provides an advantage to any legacy operator in the race to introduce 5G, and subsequently to be able to focus their efforts and capital on a smaller (although not insignificant) investment in the 5G New Radio rather than simultaneously having to invest in a new 5G core. There may be a need to update the software and possibly the systems of the LTE core, but that cost is relatively small compared to adding an entirely new core to support 5G.

The 3GPP specification provides for six options for the 5G deployment, numbered 1–5 and 7. The general view is that option 2 for Standalone and option 3 for Non-Standalone are preferred for deployments, although the specifications require that all options be supported. The six options are shown in Figure 6.3.

In option 3, the 5G handset will establish control and signaling with both a 4G and 5G base station. The 5G base station will be capable of providing the eMBB features of 5G but is not likely to meet the intended capabilities for URLLC or mMTC. Traditional voice calls would still flow over the 4G LTE

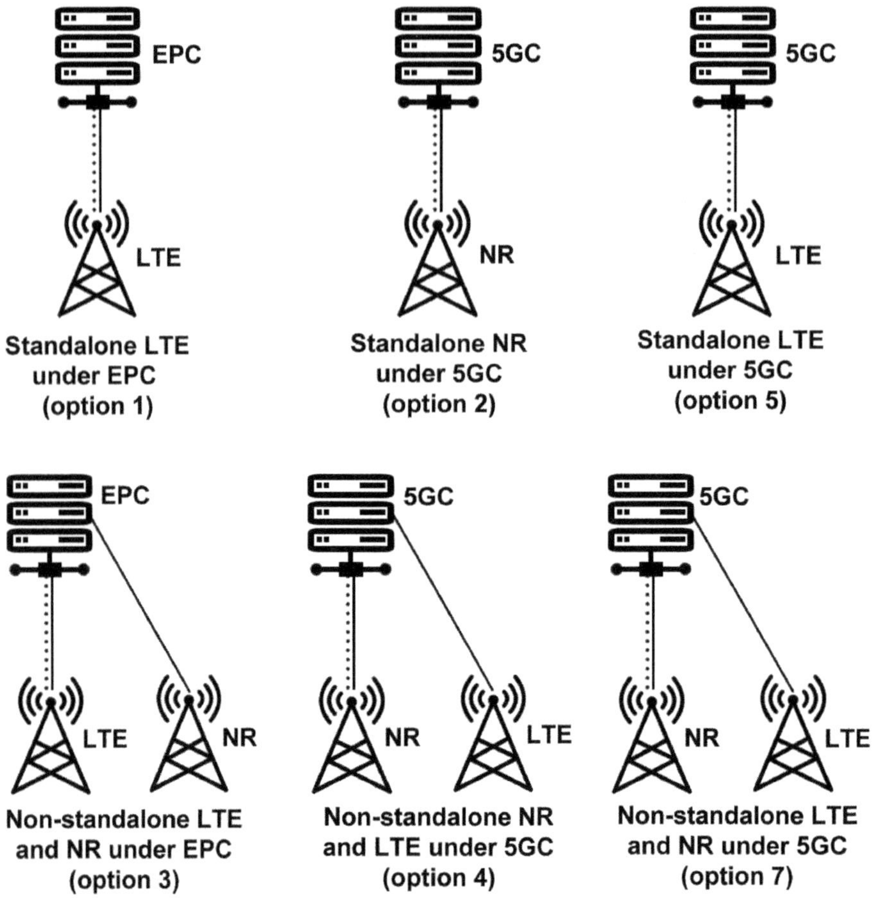


**Figure 6.3** Six options to deploy 5G SA and NSA.

network exclusively, and all other data would flow over the 5G RAN. The LTE RAN, also called eNodeB, is the primary or controlling node for the 5G RAN, while the gNodeB (3GPP-compliant implementation of the 5G-NR base station) are secondary nodes.

Option 3 has three configurations, what we will call standard option 3, option 3a, and option 3x. 5G deployments using option 3 may also support 4G LTE from the UE. As was seen in Chapter 2, many UEs support multiple generations of the RAN; the difference here is that when a UE is connecting to the 5G RAN the 5G core is also fully engaged in the communications for data and control. This option would fully support all the capabilities and use cases of not only eMBB but also URLLC and mMTC, and potentially, in addition, network slicing.

The success of this multiple-option introduction—allowing existing operators with legacy 4G networks to introduce the latest G of RAN while reusing the previous generation of the core network, and pure next generation RAN and core for other operators—is a model that has market and cost benefits and will likely be modeled again B5G.

## 6.6 Private LTE and Private 5G

One emerging trend seen in enterprises is an interest in deploying and operating private cellular networks, in particular private 4G LTE and 5G. There are several factors contributing to this growth, one is simply the availability of spectrum that is not directly licensed to the CSPs in some jurisdictions. Another is the possibility for network operators to offer slices of their network and to allocate some of their licensed spectrum to specific users. These two spectral options are fundamental and necessary yet not sufficient for some deployments; the enterprises have business needs that are creating the pull as they begin to introduce Industry 4.0 [7] into their production networks. Much of this is being driven by the need to bring compute resources closer to the point of data origin and to act on that data closer to the source. This is often coupled with discussions around edge computing. Many enterprises have operated wireless (Wi-Fi and other radios) in their production environments for a significant amount of time, and indeed these solutions are likely to continue in a hybrid form working in conjunction with private wireless in the enterprises. Since this is the case, it is useful to consider why private LTE or 5G wireless are being considered. One reason is simplification of the operations and the ability to more easily support mobility of devices; both machines and people moving around a large factory floor for example. The usage of a SIM (physical or electronic) in a UE ensures that there is no complex registration necessary to join a closed network wirelessly. This and security features found in the mobile network can be a significant differentiator for an enterprise. A second reason is the ability to easily shape traffic; that is, keep local traffic local and private. Sometimes this concern arises to protect sensitive personal information as might be found in health care, and possibly in an industrial setting where the enterprise may have no desire to let their production data leave a plant for any number of business reasons. Finally, as mentioned earlier, much of the operational data found today simply falls on the factory floor because there is no capacity to transport the data back to a central location for storage, analysis, or any other action, due in part by the volume of the data and possibly due to latency concerns. Edge computing at the factory or enterprise location opens the possibility of shunting the data close to the source and making analytical decisions in near real time.

While the capacity of 5G brings the possibility of UEs with significantly greater capabilities, not all applications have a need for the capacity of the 5G network. If this is the case, there is a logical question that can be asked: Could 4G or even 3G suffice and if so, why is 5G the desired technology? These are all reasonable questions, particularly when the capacity of the device does not need the eMBB, URLLC, or mMTC capacity of 5G for example. Two leading reasons are offered; the first has already been mentioned, today spectrum is being made available in some jurisdictions that allow an enterprise to operate in the 5G (or in some cases the 4G) spectral space without the expensive licenses incurred by the operators. Second, the maturation of the base stations has developed to a point where the return on investment (ROI) of deploying a base station does not require the same level of ARPU typically found in the commercial CSP network. Initial 3G base stations were sized to meet the needs of a CSP to support thousands of subscribers per base station, and while the technology may have existed to build microbase stations there was simply no market need. Since the early days of 4G, small cells have been available and continue to mature to the point where modestly sized 4G and 5G base stations are now available that do not require a multimillion USD per year return on investment.

One study claims that there are four quadrants for the deployments of private wireless. These are those led by an NEP or other supplier like a system integrator to the CSP, led by the enterprise itself, led by a CSP or finally, an alternative service provider (e.g., a hyperscaler). Omdia [8] in a private analysis report provides an insight where the CSPs in this area have between 20% and 25% of the market, with the first two, the NEPs and SIs, and the enterprises themselves having over 70% of the market share. It appears to be clear from this report that the CSPs are not leading in the deployment of private 5G at this time. There may be several factors to consider here. CSPs are not interested in this market. This can easily be discounted. The market to justify a 5G investment required by the CSPs cannot rely on an increased ARPU from the consumer market; thus, the market for 5G users have to include enterprises and their nonhuman use devices. This leads to the question: Why are the CSPs behind in private networking? We will leave this question at this time with the following suggestion that this is an interesting business case study that many are pursuing and there are multiple factors and possibly multiple reasons.

## 6.7 The Cost of 4G and 5G Is Changing

The market is changing for providers of the network elements (the network equipment providers). Nodes that in the past had multimillion USD price tags are decreasing in cost in part due to the ever-increasing march of technology, and possibly due to the entrance of disruptors into what has previously been a nearly closed ecosystem.

### 6.7.1 Regulatory Considerations

One of the advantages of disaggregation is that vendors are not solely focused on developing products to be consumed by an ecosystem where only large-scale deployments of RAN and core elements have a viable business model. This has been the model in the past, in part due to the licensing of spectrum from the federal governments where the CSPs have spent significant sums to obtain licenses (billions of USD in many cases) and therefore built a business model that generates a return on that investment where only large-scale deployments were viable. The emergence of small cells in the 4G era and in some jurisdictions the introduction of "lightly licensed" spectrum—such as the CBRS in the United States, where the CSPs were restricted from bidding—have opened up markets to new deployments where large-scale solutions are no longer required. Other nations are following with the introduction of spectrum that is well suited for non-CSP usage by enterprises large and small. Large factories, ports, and rail locations that are people sparce and resource dense are ideal candidates for the use of managed spectrum in a private or semiprivate wireless network that relies on 5G or earlier standards-based technology. These enterprises may have used a variety of unlicensed spectrum solutions in the past and may continue to do so in the future in conjunction with the newly deployed 5G. There are several advantages to using the 3GPP-based solutions, including that they are now affordable and there is improved manageability of the services and security of the network in addition to improved mobility. Devices enabled with a SIM (or eSIM) where the SIM is associated with the network allow devices to securely connect to the network without any additional registration on the part of the device user. In addition, the range of the radio connection can, in some cases, be significantly greater when using licensed spectrum over the case where unlicensed spectrum is used. There are two reasons for this: (1) licensed spectrum in many cases allows the access point (base station) to operate at a higher power level and as a direct result have a greater range of operations, and (2) licensed spectrum has less interference from other users. 3GPP has spent decades improving the mobility capabilities defined in the standards and vendors have created devices that are designed to seamlessly roam from cell to cell without user intervention. There are several use cases where devices that are part of a machine or an autonomous system may have relatively limited mobility but can coexist on the same radio network devices with a wide range of mobility within the geography of the enterprise. There are several enterprise use cases emerging that span beyond the traditional cellular network use cases. One example is in the energy utility sector where power gird substations may have improved connectivity using 5G. This enables field maintenance engineers in remote areas to have access to a mobile network along with devices for monitoring the operations of the substation. An example would include video surveillance for hot spots on equipment and for physical security purposes. The bandwidth needed for multiple surveillance cameras fits well into the 5G capacity, and a mobile worker could roam onto the energy utility network independently of any specific carrier coverage that may not be available in remote locations.

Initiatives to boost vendor competition, for example Open RAN, have varying degrees of support and involvement from the largest incumbent vendors. The largest incumbent vendors will participate at an appropriate level that does not cause a negative impact to their existing product roadmaps and ongoing revenue streams, and at the same time be positioned to pivot with a number of possible approaches should technology adoption reach an inflection point. One of these approaches takes a page from Cisco in the pre-Y2K era (year 2000 device clock transition). Cisco was very adept at acquiring innovative technology start-ups and rapidly and effectively integrating their technology into the Cisco family of products or removing a potential competitor and placing the intellectual property on the shelf for future use. This buy-vs-build strategy can both manage the investment risk and allow for existing product teams to continue to deliver solid results on existing product roadmaps. A riskier approach is to build competing teams within a single organizational structure. This path can lead to success, but often internal conflicts, including the innovation antibodies mentioned earlier, outweigh the possible advantages. This may be why it is more common to see transformative innovation, even in the CSP market, come from outside the existing largest legacy providers. An exception is the cooperation seen in the standards bodies such as 3GPP where the industry has evolved from the 3Gs to 4Gs and now the 5Gs with an eye turning rapidly to beyond 5G. One may argue that in this case the innovation is linear and not one of transformation. We will leave this last point for others to ponder further.

## 6.8 Security in the Disaggregated Network

Security is an area of constant concern for network operators. Going back over six decades the introduction of the first modems onto the PSTN network brought security concerns for the operators that have only grown with each generation of technological advancement. The security concerns of a disaggregated system and in particular those where the support teams are using a telco DevOps model require that the CSP be nimble enough to respond quickly to the release of security patches to systems. This includes core operating systems, platform firmware (e.g., BIOS), middleware, and application-layer software, often from multiple vendors for a single platform. Today, it is well understood that no system can be totally secure and the best security practices require that patches too close security holes are applied as soon as they become available. This model can be challenging for legacy CSPs who follow a cadence of upgrades, possibly only a few each year. In addition, the model of a single vendor responsibility per node in the network also needs to be relegated to the past, which is another significant challenge for many CSPs.

## 6.9 Transforming Operations: A Use Case Example

AT&T entered into an agreement with Microsoft in mid-2021 where Microsoft would be responsible for the operation of the 5G mobile network. This included the transition of employees from AT&T to Microsoft. One of the stated goals was to help AT&T increase their competitive advantage by streamlining their operations and bringing service differentiation. This effort is to begin with the mobile core initially with the stated goal of increasing productivity and cost efficiency. This expansion into the operational domain by Microsoft complements their acquisition of Affirmed Networks and Metaswitch in 2020.

The Global System for Mobile Communications Association (GSMA) estimates that between 2021 and 2025 the global telco market CapEx will equal approximately 900 billion USD, with up to 80% being spent on 5G introduction and expansion. Operators will very likely be watching the AT&T and Microsoft partnership progress while also considering their own approach to cloudification to save on hardware and operational costs. Additional goals include improved automation, access to better data analytics, and increased ability to manage and resolve failures in near-real time while quickly responding to traffic demands. Aspirational goals include innovating new services and applications using AI and ML.

## 6.10 Beyond 5G Market Drivers

Looking past 5G, 6G is already being talked about in earnest at various companies, including Nokia, Ericsson, and QUALCOMM, and in academic circles, including MIT. 6G will require three-way cooperation between governments, academia, and industry in the early formative stages. Standardization is likely to begin in 2024 or shortly thereafter and reach maturity in the 2028-to-2030 timeframe with initial commercial launch shortly thereafter. 3GPP is planning the initial 6G specifications to appear on Release 21 scheduled for 2028. While it is too early to speculate on the functions that will drive 6G, several possibilities are emerging. The first is utilization of even higher frequency spectrum above the 5G millimeter-wave, which will lead to even faster speeds. This may come at the cost of reduced range between the UE and the RAN. Several other emerging areas are zero-energy, where endpoint IoT devices would extract their operating power from the environment or from RF. No external power source or battery would be required in the device. Improved operations and efficiency using AI and ML even closer to the RAN nodes is also being considered.

## References

[1] GSMA Option 3, https://www.gsma.com/futurenetworks/wiki/5g-implementation-guidelines/.

[2] Rommer, S., P. Hedman, M. Olsson, L. Frid, S. Sultana, and C. Mulligan, *5G Core Networks Powering Digitalization,* London: Academic Press, 2020.

[3] John, W., G. Marchetto, F. Nemeth, et al., "Service Provider DevOps," *IEEE Communications Magazine,* Vol. 55, No. 1, January 2017, pp. 204–211, doi: 10.1109/MCOM.2017.1500803CM.

[4] https://about.att.com/story/2021/att_microsoft_azure.html.

[5] ETSI GR NFV-REL 012 v1.1.1, https://www.etsi.org/deliver/etsi_gr/NFV-REL/001_099/012/01.01.01_60/gr_NFV-REL012v010101p.pdf.

[6] 5G Implementation Guidelines: SA Option 2, https://www.gsma.com/futurenetworks/wp-content/uploads/2020/06/5G-SA-Option-2-ImplementationGuideline-v1.3.pdf.

[7] SAP, "What Is Industry 4.0?" https://www.sap.com/insights/what-is-industry-4-0.html.

[8] OMDIA, *Industries and Enterprises Are Ready to Reap the Benefits of 5G,* 2022, https://tmt.knect365.com/uploads/2646-Omdia-BearingPoint-5G-EBook-V1O-fed6e558e89a-f5a61a117025401bd4b2.pdf.

# 7

# Designing Virtualized RAN

## 7.1 Virtualizing the 5G RAN

This chapter covers at a basic level the design of the modern virtualized RAN, in particular the 5G RAN, and the standards from 3GPP, which is the main standards body. In addition, the O-RAN Alliance and virtualization of the RAN (vRAN) design principles will be discussed. Splits in the design will also be covered in this chapter as it applies to the protocols that allow this separation to be implemented.

5G has several standards bodies, 3GPP being the leading body with others contributing their share. We will provide a review of the current standards (not in detail, as that scope alone would fill a complete work), and the importance and working of the standards bodies is discussed.

One of these bodies is OpenRAN. The OpenRAN alliance builds on the work of 3GPP and provides a truly open specification for the interconnection of the various RAN elements. O-RAN is an effort, not really a standard, and it should not be confused with OpenRAN (but it is often confused) or overloaded.

vRAN is a concept that complements these efforts and standards and is different from the above discussions. Often it is found that a vRAN is open, but this is not necessarily a requirement. Here the various considerations of hardware and acceleration in a virtualized 5G RAN will be discussed. The vRAN ideally is a decomposition of the RAN and the complete realization may require that all the elements are sources from one vendor, where the OpenRAN concept and standards work specifically allows for the composition of the RAN to be sourced from multiple vendors, and it may not be virtualized (in theory). In practice OpenRAN implementations are virtualized.

Radio splits are various models to deploy a vRAN, and many CSPs will have different requirements on power, space, latency, coverage, or other considerations that drive different splits, such as 7.2 or split 3 as they are called by the standards (do not confuse this 7.x split with this chapter's "7" numbering).

### 7.1.1 It All Begins with the Standards

3GPP in Release 15 provided the introductory model for the design of a virtualized RAN for 5G. Here we find the gNB being introduced; in 4G it was an enhanced Node B (eNB). Subsequent releases by 3GPP have advanced this work and the current standards are at Release 18. The gNB is the 5G radio and associated systems necessary to realize 5G in a network. In this design there are three significant elements, shown in Figure 7.1.

These elements are the remote radio unit (RRU), the distributed unit (DU), and the centralized unit (CU). The standards created a separation of functions that are called split points during the initial phases of the standards development (see Figure 7.2), these are "Options 1-8" where Option 8 is closest to the RF, and Option 1 is closest to the core of the network. Only split options points 2 and 7 matured in the finalization of the standard, and as a consequence, have been realized in operational systems. The interface between the RRU and the DU is physically carried over what is termed the fronthaul and is represented by Split 7 and 2x. From the DU to the CU, the interface is called the midhaul and is Split 2. From the CU to the core network elements there is not an associated split, and this interface is often referred to as the backhaul. The interfaces between the DU, CU, and mobile core can be virtual interfaces; that is, while these modules are described separately, they in fact can be realized on the same systems or a collection of systems, and in some cases

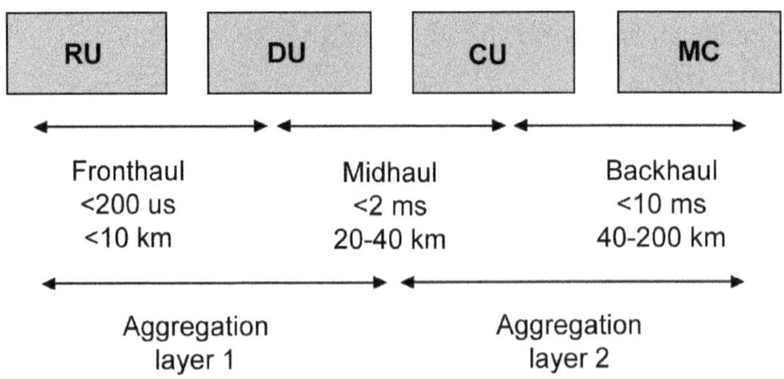


**Figure 7.1** 3GPP Release 15; the gNB modules.

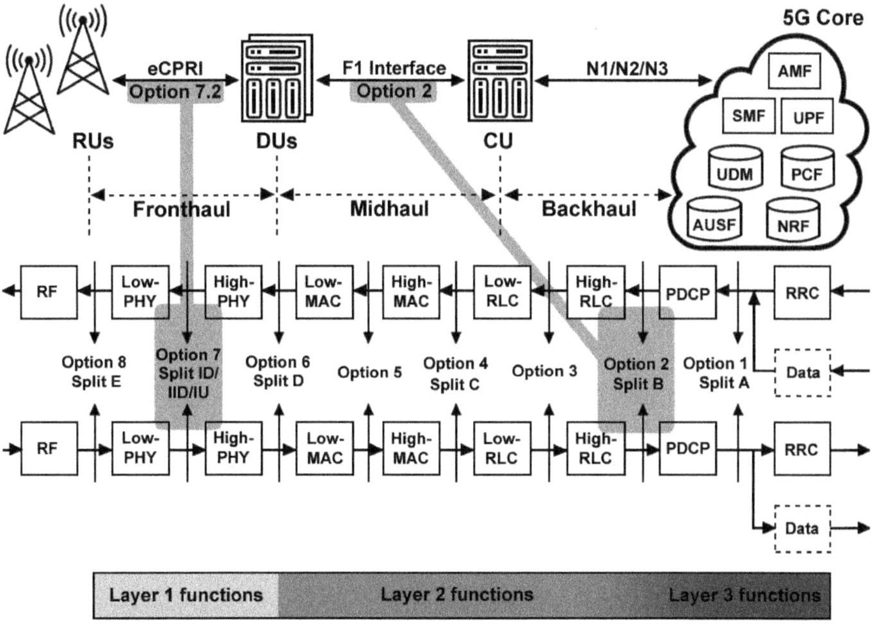


**Figure 7.2** The splits in the 5G.

there may be only internal software realizing the interface between functions. Another way of saying this is although they are called interfaces, they may in fact be implemented (realized) as software-only interfaces, where there is not a physical device (wire, fiber, or other medium) connecting the elements. This is one of the key features found in the cloud-native design embedded into the 5G specifications.

This a very broad range of deployment options, and for the designers and developers to specify how the modules interoperate. As a direct result the lower layers of the protocol stack are part of and provided by the host operating system available.

### 7.1.2 Operating Systems of Choice

With very few exceptions the operating system of choice is a derivative of Linux; some consider VMWare and Wind River as having enough UNIX in their legacy for us not to quibble here. While there are some examples of development and deployments on "native" Linux, commercial efforts are using real-time enhancements in the kernel at least for the DU. In general, the native kernel today is not capable of the guaranteed response time required for the packet processing in the DU and in some cases in the CU. The software company Wind River developed a real-time operating system (RTOS) from their inception, which has been proven in commercial deployments of vRAN. The now deprecated CentOS was one of the earlier instances that supported the real-time features necessary in the kernel. More recently, Red Hat and VMWare have developed instances capable of being configured for real-time performance. The essential part is guaranteed response to inbound packet processing and the ability of the packet-processing software driver to have run to completion in critical sections of code.

We will allow others to debate whether VMWare and Wind River are, or are not, derivatives of Linux. Clearly, they all share at least some constructs that can be traced to the origins of UNIX, if not in code, then at a minimum in the base design and theory of operations, as in all I/O is file system based, for example. We have yet to see RTOS implementations realized on the Microsoft or Apple operating systems. Over time this may change, but for now, sticking with those derived or based on classical UNIX would be recommended. In addition, when discussing the core operating system, we often either intentionally or possibly at times without regard for technical rigor, discuss the kernel of the OS and the upper layers of the operating system as one entity. There are cases where various enhancements (and in some possible cases "enhancements" may have actually been removal of some functionality) have been implanted in the kernel of the OS, such as implementation of real-time features.

### 7.1.3 Supplementation of the OS

There are cases where a non-Linux OS may be used as part of the protocol stack. If so, it is reserved for the lower layers, while the upper layers are provided by the OS vendor. The lowest layer may be specific to the physical media, for example between the RRU and the DU, but all other components of the protocol stack are part of the operating system. Figure 7.3 shows the high-level abstraction of this protocol stack. The developer of the upper layer application layer can implement their workload relying on the OS to provide the lower protocols (e.g., JSON from IETF RFC 8259 [1] running on top of HTTP/2, which can be encrypted using TLS). Transport Control Protocol (TCP) and IP that form the foundation of today's internet (thank you Vint Cerf and others). These are well proven technologies providing reliable transfers along with flow and error control. Beyond 5G there are discussions going on to replace TCP in the future with Stream Control Transmission Protocol (SCTP), which is a significant enhancement to classical TCP that has been around for nearly two decades but has yet to totally replace TCP in any large-scale CSP deployment. Interestingly, SCTP was originally developed to be used to carry SS7 (Signaling System 7) over internet connections [2]. SS7 was a foundational protocol in the 2G mobile network. We find it interesting that echoes of past protocols could reverberate into the next generation of the mobile network. One reason to con-

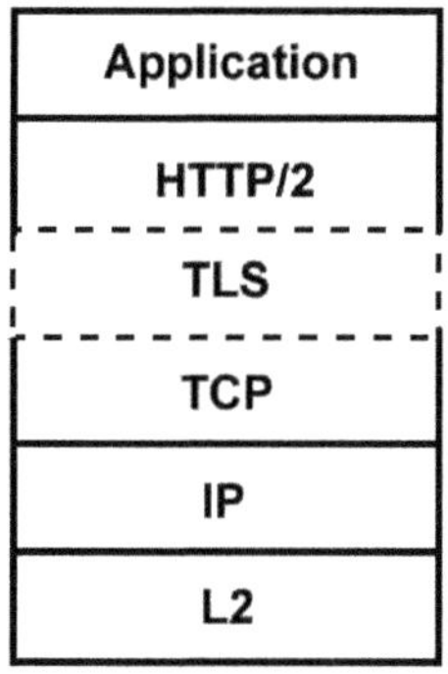


**Figure 7.3** 5G Interface Protocol stack.

sider SCTP is the native ability to support multihoming and redundant paths, thus freeing the developer from even more lower-layer protocol development work. An additional advantage to using protocols found natively in the Linux and other commercial OSs that are in the toolset and skill set of the maintenance staff at the operator is to leverage a wider base of common knowledge. These protocols all have a common base or foundation and few proprietary tools. This leads to reduced training requirements and lower overall operational cost in addition to the wider deployment base outside CSPs, which tends to remove defects quicker due to the fact that many more users engaged in the software development.

## 7.2 The Continuing Evolution of the Standards

Readers of the Release 15 [3] will soon see that the specification actually labels the names of the RU, DU, CU, and MC differently than what is in common use today (and used here); the specifications tag these as NG-RAN, gNB-DU, gNB-DU, and 5GC respectively. (At the time of this writing Release 16 is final [4] and Release 17 [5] is in the final reviews with work on Release 18 well underway.) Possibly an additional point of confusion in our collection of abbreviated node names will come when we discuss the eCPRI protocol later, as the DU and CU are renamed again. As with much of the trends in the CSP industry, a shorthand has been introduced and as such, RU, DU, CU, and MC have become the popular nomenclature. The careful reader of Release 15 will notice that the actual document is called "3GPP TR 21.915 version 15.0.0 Release 15," but it is simply referred to as Rel 15. From Rel 15 we also quickly find that this document is far from stand-alone as the reader is also directed to 3GPP TR 21.915 version 15.0.0 Release 15 [6].

The CU is further separated into control plane (CP) and user plane (UP) functions in the specs; these are shown in Figures 7.4 and Figure 7.5,

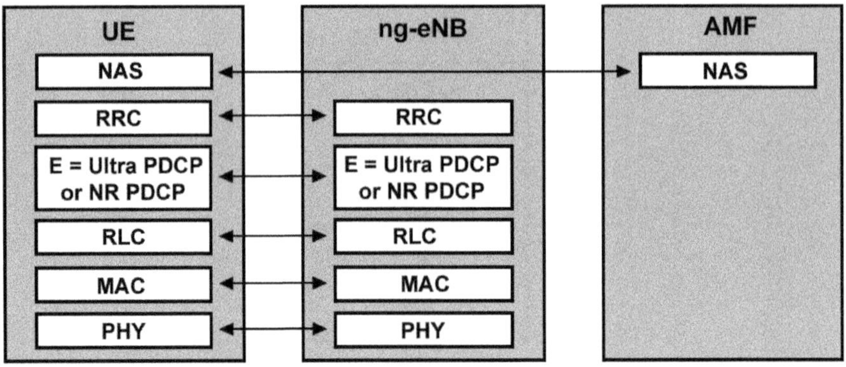


**Figure 7.4** Control Plane Protocol stack.

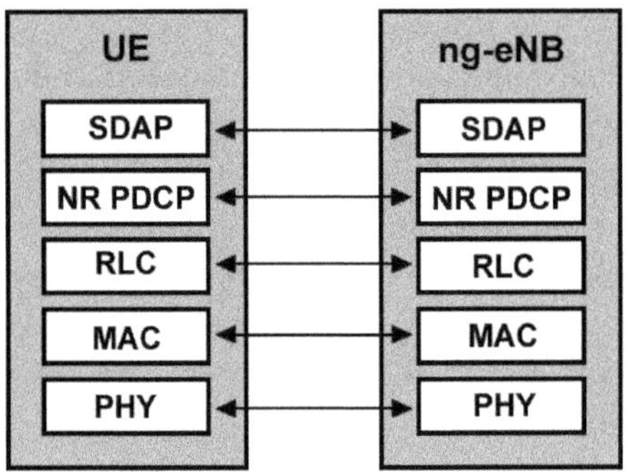


**Figure 7.5** User Plane Protocol stack.

respectively. The CU-CP may connect to multiple CU-UPs, which enables the implementation to scale the user plane traffic independently of the control plane, part of what was earlier termed CUPS.

## 7.3 Attaching the UE to a Network

The above specification is relatively short but contains specific and detailed messaging procedures in the gNB-DU and CU flows. While our attention is on the RAN, the real purpose is to allow the end point devices (the UEs) to do what they do best. To this end the RAN specifications capture the protocol to bind the UE to the network and allow for mobility. We find this in the discussion around the initial access of the UE to the network, sometimes called attachment. Also included are the protocols for specific mobility cases including intra-CU mobility, which consists of inter-DU mobility. In this case, the UE moves from one DU to another where both DUs are connected to the same CU. Redundancy, or resiliency as it is described in the specification, is supported as an option (but one should always design in redundancy if the product is going to be accepted in a commercial production network), where a DU and CU-UP may have connections to multiple CU control plane nodes. In addition, a DU may connect to multiple CU user planes (provided the redundant CU-UPs are controlled by the same CU-CP). The specification further requires that the CU-CP is responsible for establishing the DU to CU-UP connections. Furthermore, the CU-CP is responsible for the selection of the CU-UP to support any services requested by the UE.

There is a binding of a specific UE when the UE device establishes a new logical connection to the RAN and is associated with the AMF. The application protocol identity (AP ID) is unique for the logical connection between the UE and references on the Xn, F1, and E1 interfaces for this RAN node. The RAN node association is also present in the signaling. (In theory a different carrier may also assign the same AP ID as there is no control association between the two carriers to prevent this, but this occurrence is not a concern as the AP ID has local carrier scope only.) The use of the AP ID is bidirectional; a sending node includes their AP ID to a receiving node, which stores (remembers) this ID for use for the duration of any connection. Similarly, the receiving node will include an AP ID of its own in any returned messages. These two AP IDs provide a logical association that identifies the connection, which is similar to the use of ports between sending and receiver in IP protocols. The two ends of the connections will continue to include both AP IDs in every message between the two for the duration of the connection. There are implications to this uniqueness when mobility causes the UE to be associated with a new RAN node. The specification provides the description for the transition from one RAN node to another RAN node such that the UE does not have to generate a new association to the AMF (into the core network). Additional associations are specified for the Xn interface for dual connectivity between the RAN and the UE. The CU and DU to UE over the F1 interface, likewise, requires a unique ID, the F1AP ID, and again this ID is unique and maintained for the duration of the association (logical connection). There is also a uniquely configured ID between the DU and CU (the DU provides this ID) for use in the F1 setup procedure between these two nodes. That is, if the either the CU or DU start (initially) or restart for any reason, there is an association between these two that is not associated directly with a UE, rather only for use for signaling between themselves. Reference [7] provides the structure for the UDP messages along with the identification of the source and destination ports for this association.

### 7.3.1 The Roaming UE

There are several points in the network where a mobile device that is moving causes a transfer from one RAN cell to another and must be handed off. The roaming device can also be reassociated with a different RRU, DU, or even CU as a result of its movement or roaming. If the device moves from the coverage area of the carrier's network, a reregistration onto a different network is triggered or once back in coverage range of the original network it will revert. These would appear to the network as initial connections rather than a handoff. There are several handoffs that are discussed in this portion of the specification; one is the intra-DU handover. In this case the UE roams far enough that the serving DU is no longer the desired DU, yet the current CU remains the active CU.

### 7.3.2 The UE Detailed Signaling Flow

When a UE first (under a variety of conditions) established contact with the gNB the signaling flow below is used. Here we see that the DU, CU, and AMF all have knowledge of the UE at the completion of the initial access procedure (Figure 7.6).

Once the UE has completed the initial association it may move from one DU to another, either because of actual location change of the UE or because of a failed air interface between the UE and the RRU connected to the initial DU. The UE sends regular measurement reports to the DU and the DU may report to the CU the need to transfer the UE to another DU. If this occurs the CU will inform the target DU, and if the target DU can accept the UE, it will reply back to the CU acknowledging its ability to accommodate the UE. The initial DU will notify the UE that it is going to be switching to a different DU with a Connection Reconfiguration message. The UE and the newly targeted DU will complete their transition, and the target DU will inform the CU once this has been successfully completed. The CU will then inform the original DU to release the connection to the UE. If there were packets in transient that failed to reach the UE during this transition, the CU will forward them (either before completion of the transfer or after; it's up to the implementor) to the newly anointed DU.

Another transfer can occur where the UE moves from one cell on the RRU to another cell on the same RRU, and hence remains within the same DU. This case is called the intra-DU handover. This handover is not as complicated but nevertheless requires some treatment. The CU initiates this handover by providing the UE with a new uplink ID (through the DU), and the DU provides a new downlink ID to the CU. The DU and CU will continue to use the old IDs until the buffers are exhausted or the reestablishment is complete as the UE is likely still in range of the original cell for a short time while moving into the range of the new cell. These cells are very likely on the same tower

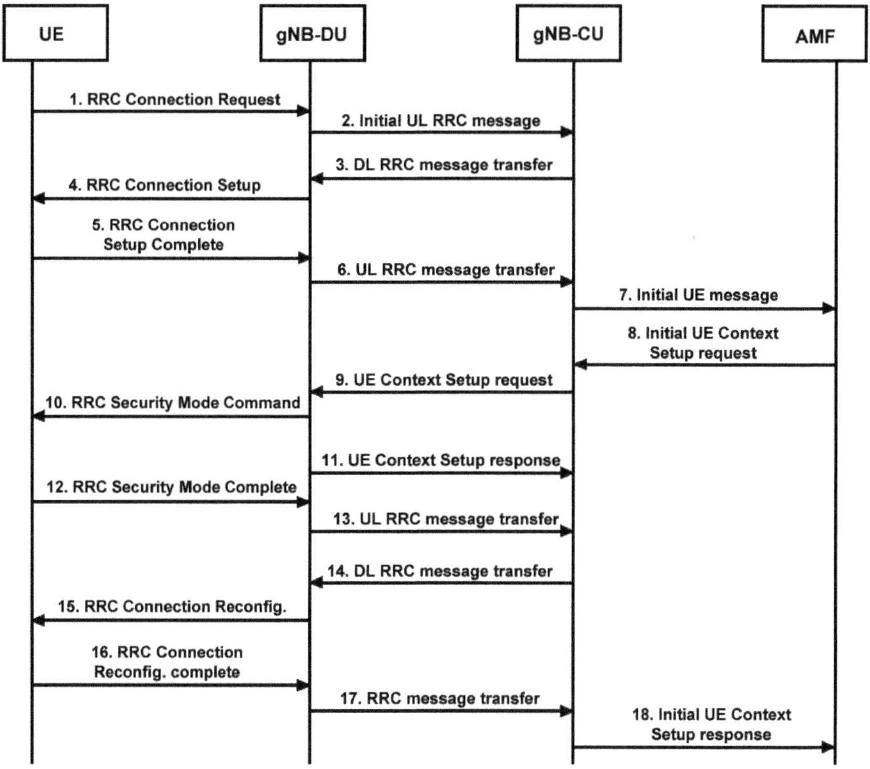


**Figure 7.6** UE initial access procedure.

(typically an RRU is not split to different physical locations), and in fact they may be different antenna segments on the same larger antenna or they may be on one of the adjacent antenna segments. Additional inter-DU mobility cases exist but are not covered here.

One of the salient features of the mobile network and in particular of the 5G network is its resilience (the ability to recovery from errors autonomously). We will see in a later chapter on the RRU a number of error recovery procedures (relating to dropped packets, etc.). At this time, we should consider the interaction between the CU and DU to recover from lost or dropped packets. The CU has an ability to retransmit Packet Data Convergence Protocol (PDCP) data units (PDUs). In the case where the DU determines that the radio link to the UE has failed or is failing (in this case the UE cannot report a measurement to the DU to initiate a handover as was discussed previously, the DU will signal to the CU declaring a radio link outage. If the CU is able to locate another DU that is able to serve the UE, the CU may retransmit the previously undelivered data units to the newly chosen DU without effecting a handover. It is anticipated that the original DU will see the radio link to the UE return in short order, and when that occurs it will report the restoration to the CU, which will then resume sending the data units to the original DU. When one is considering the upper-layer software that would realize these handovers and the error procedures that are taking place, a well-designed and nearly error-free implementation may be realized that accurately implements these protocols. This implementation could be considered incomplete when viewed from the operational lens of a CSP. While the specifications make no mention of logging and counting these events, production systems that silently comply without any logging or tracing ability (via peg counts and sliding windows or other methods) would likely fail to be adopted in any real-world production environment. Developers may be inclined to view the specifications as containing the complete set of requirements; after all they clearly use the shall/should/may found in requirements documents. Our considered guidance is that these specifications document the minimum set of criteria, or, as one might find in a mathematical text, as meeting the necessary conditions but not the sufficiency criteria. Imagine a “flapping” condition where one DU is continuously reporting link outage and recovery to a CU. This may be an indication of some physical impairment that merits attention. If there is no reporting, an unnecessary burden has been placed upon the operational nodes that may lead to some other degradation of the service over time or increased load on the CU, increasing its power consumption and reducing its overall capacity. Finding and eliminating these squeaky hinges before they manifest as real problems in a live network is the hallmark of great products in the CSP network [8].

## 7.4 Initialization of the DU to CU Connection

Prior to any UE establishing the initial access procedure to the DU, the DU must establish an association with the CU and the 5GC. At provisioning time, the DU will be configured (this will be vendor-specific in many cases and with elements from different vendors, e.g., the OpenRAN DU will of necessity extensively validate this prior to production rollout) with data on the specific cells under its control. The DU will initiate connectivity toward the CU, which it will also have been configured to know about, with a Setup Request message over the F1 interface. This message will include information about the cells that the DU has configured. The CU will communicate this information deeper into the 5G Core and the reply back to the DU may include a list of cells to activate. This messaging will be reused in the case of a cell failure later. The DU will then reply back to the CU informing the CU which cells (hopefully all of them!) have been activated. The CU will then have knowledge about which cells can be used for any UEs that complete the initial association through this DU. The CU will delete from its list any cell that the DU fails to activate during this initial procedure. Clearly this implies that the CU has knowledge of the intended cells (a provisioning function) that the specific DU has control over.

The CU and DU will each maintain a state for the known cells, which are Active and Inactive. A cell that is in the Active state is able to serve any UE, and a cell in the Inactive state is prohibited from serving a UE. The CU as part of this initial association with the DU may also establish connectivity to either (it could be both) another gNB or an LTE eNB. This level of redundancy provides additional handoff options for a connected UE.

### 7.4.1 Back to the UE Attachment

The first thing a UE will attempt is to establish a connection to a gNB; the RAN portions will be discussed in later chapters. Here, the UE has found a suitable gNB and initiates the procedures to connect to the gNB. The UE initiates the connection by sending an RRC connection request message to the DU, which will send an Uplink Initial RRC to the CU control plane (CU-CP; the CU data plane connection will also be established) later in the procedure. The CU-CP responds to the DU with a Downlink RRC message transfer, which causes the DU to reply to the UE with a connection setup response. The UE will acknowledge back to the DU with a Connection Setup complete message.

## 7.5 The 80/20 Rule

Many of the procedures described above are sunny-day procedures; that is, when everything goes well. There is limited discussion on the implementation of retires or conditions where messages may be missed (or too much time elapses between sending and receiving). Even with the error recovery procedures in the lower layers of the protocol stacks, errors may and do occur. These error conditions often require more development effort (and event logging and tracking) than the sunny-day procedures would imply. This has often been described by experienced development engineers as the 80/20 rule (while not really a rule but rather a large-scale trend). The rule works this way: 80% of the functionality is often achieved within the first 20% of the development effort, while the remaining 20% of the functionality (our view is that this is actually an asymptote, as software is never "done") will require an additional 80% of the time remaining or available.

One of the other issues that has been encountered is the use of reference software from open source or from other sources. This can significantly complicate the development effort where these modules are not standard library components, but rather are intended to be inline integrated source code. The challenge here is to maintain a single-source code control structure, where from time to time a new release of the reference software becomes available. In this case, the original and updated source code configurations must be merged. While there are very high-quality tools available, the deviation from release to release can add an indeterminate impact on the release schedule as well as introducing new bugs. While no perfect solution exists to this problem, the need to continuously practice good development hygiene on both sides with well controlled change control practices may serve to ease the burden.

## 7.6 Splitting the RAN: Revisited

The RAN splits are functional separations, with the RRU responsible for the RF processing and the functional processing below split 7. The DU is responsible for the functional processing above split 7 and below split 2. Finally, the CU is responsible for the functional processing above split 2.

There are several architectural advantages to the design laid out by 3GPP. The first is that the DU and CU can be deployed in a purely distributed design. This implies that the DU and CU can be optionally colocated and even resident as separate functions on the same server or servers, while there are other cases where the DU can be colocated with an RRU. Similarly, a CU may also be colocated with the RRU (when the DU is also colocated). There is often a many-to-one relationship where multiple RRUs are associated with a single (with a redundant configuration) DU, and in some cases multiple DUs may be associated with consolidated CUs. These decisions are most often based on network capacity requirements of the operator.

### 7.6.1 FEC Processing and More in the RAN

The forward error correcting processing (FEC; encode and decode) while possible on a general processer today, in the real world operation at scale is done today in special acceleration hardware. While there is work being done to implement the FEC function using new instructions in some general processors, this is still a few years out from full-scale deployment. The current state of the art relies on fixed-function ASIC, enhanced ASIC (eASIC, which is an enhanced ASIC that provides faster turnaround time in manufacturing and may include some level of field configurability), or FPGA implementations.

We will set aside the vRAN for the moment and focus on virtualization of the DU and CUs using standard high-volume servers. In many cases, these servers are supplemented with FPGA cards to improve performance, reduce latency, and address TCO concerns. Here, there is some debate on the structure of the FEC processing that in the views of many is not yet settled. One common implementation is a lookaside, where the core processing hands off packets (one or more) to the FPGA and immediately returns to continue lower- and upper-layer processing of additional packets. Then, via one of several methods, the core processing will be notified that the FPGA has completed the processing and packets are ready to be retrieved by the core processor for the remainder of the treatment on the packet. The alternative to lookaside is a blocking method where the core processing will block (very briefly) for the FPGA to complete the processing before continuing the processing of the packet. Each of these techniques has a set of advantages and disadvantages. The blocking method relies on some interrupt or other method of polling to detect the completion of the processing of the FPGA, which may complicate the synchronization and add overhead to the processing. This disadvantage in complexity may be offset by better overall performance (e.g., less hardware for the same workload when compared to the nonblocking method). The nonblocking method, while possibly much simpler to realize both at the core application layer and within the FPGA, would potentially require more CPU and FPGA resources to process the same workload. Figure 7.7 shows a typical FPGA accelerator with blocks allocated to the upstream and downstream processing.

A serial or nonblocking processing on dedicated hardware is often found with an ASIC implementation. The current state of the art using accelerators remains a lookaside design. While this design is less difficult to realize, resources are not as economically utilized as those found in a heterogeneous design.

The use of FPGA accelerator cards for the FEC processing between the DU and RRU has exposed one of the challenges of the disaggregated model. These cards are often sourced from vendors that are not the OEM/ODM of the standard high-volume server, and as a result, the dependency on them for the acceleration relies on the lifecycle availability of the cards and compatible drivers.

A second consideration is that the programming of these cards for a specific function such as the FEC is not as straightforward as programming at

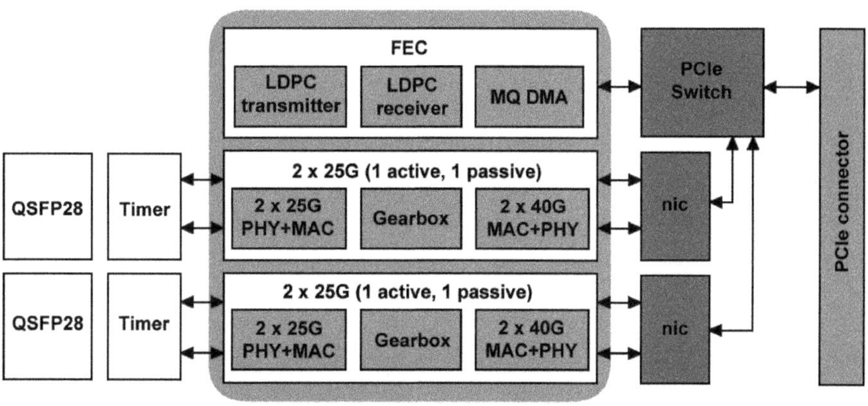


**Figure 7.7** FPGA block allocation for FEC and fronthaul I/O.

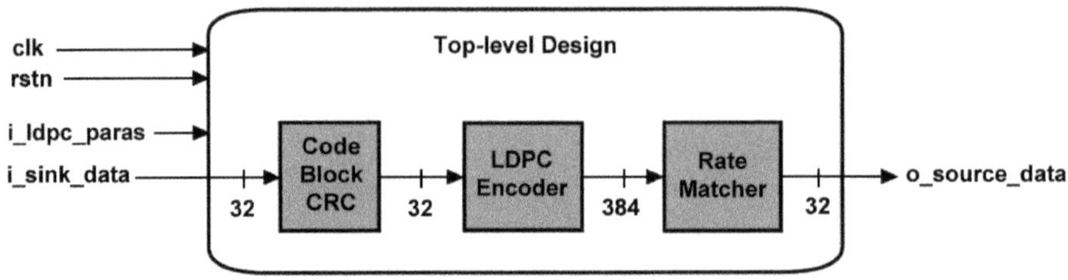


**Figure 7.8** FPGA transmitter signals.

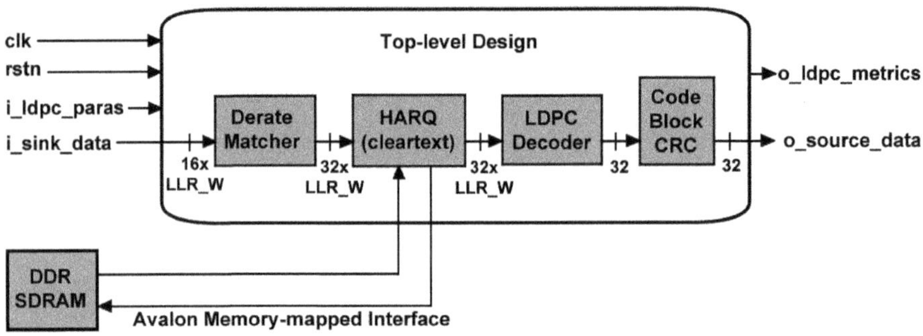


**Figure 7.9** FPGA receiver signals.

the higher layers for the server. Programming FPGAs requires a skill set that is much rarer to find in practice. At this point one might rightly be asking why then use an FPGA, and while this is a good question, we will dispatch it quickly. The nature of the FEC used in 5G is based on 16-bit integer math, and while this can be done on a 64-bit CPU core, there are complications that can degrade performance and thus challenge the cost profile of the virtualized solution. Figure 7.10 shows a functional block design of the channel coding on the FPGA. The channel coding is performed on the DU and supports the multiple transmit antenna as well as implementing the FEC and HARQ for error protection of the transmitted signals.

The FPGA can also be used for the compression and decompression of the fronthaul input and output as shown in Figure 7.11. In addition, the packet classification is included in this processing flow.

Simply put, doing 16-bit math with 64-bit registers wastes three quarters of the cycles for each calculation. The FPGA is ideal in that the logic allows for the construction of 16-bit registers and the design of application-specific math to address this performance issue provided one can source the FPGA cards. There is another solution, which is being played out now: add into the CPU architecture the ability to operate on a quadruplet of samples with 16-bit registers natively designed into the instruction set and CPU architecture. One CPU

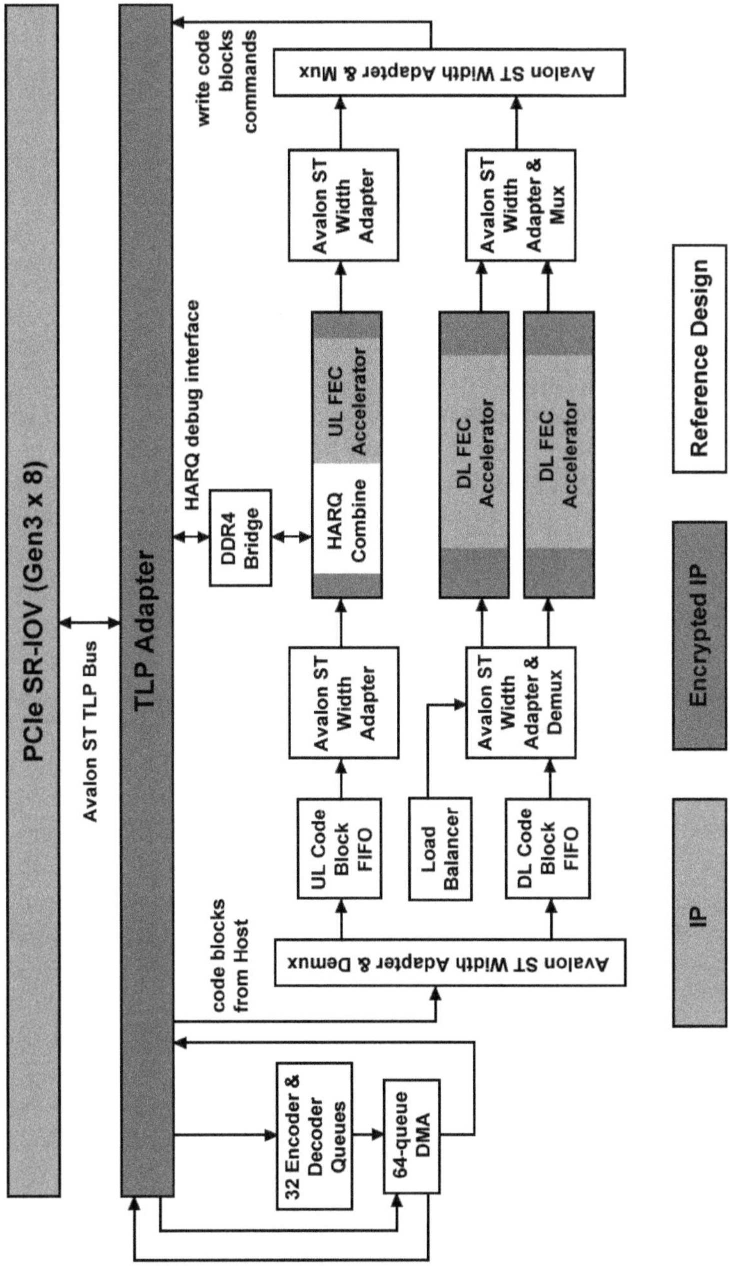


**Figure 7.10** FPGA channel coder for 5G.

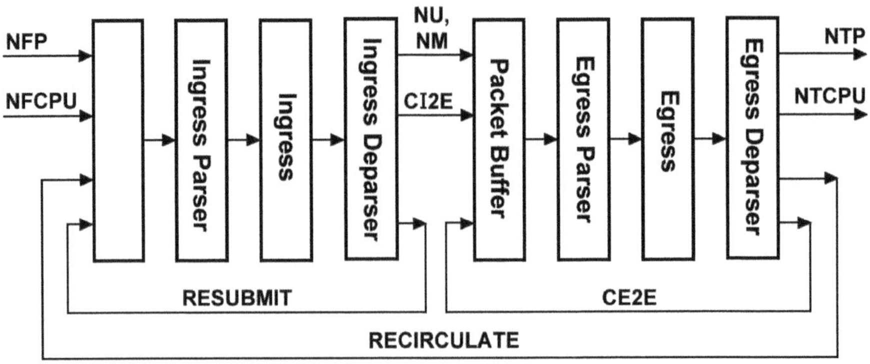


**Figure 7.11** FPGA fronthaul I/O.

vendor is adding this capability into their 5G-capable CPU and these will be available from a number of original equipment manufacturer (OEM) vendors in due course. Here again is an opportunity for the ISV community to innovate and exercise their development skills. The FPGA code is not directly compliable in a standard high-level language (for example C or C++); it is likely that some ISVs may offer library support initially, while others begin to learn to use the macro flags and architectural software designs that use these CPU features. The IEEE has specified floating point in IEEE-754-2019 [9]. This enhancement in the CPU's native capability is ideal for the signal processing required for antenna beamforming, precoding, and in the minimum mean squared error (MMSE) calculations. With traditional 64-bit registers, operation of 16-bit values, while possible, is not ideal from a performance perspective. These enhancements are realized as a collection of parallel 16-bit registers (32 in total), which can perform in parallel the calculation on 32 16-bit samples at one time. This level of processing, often called single instruction multiple data (SIMD) is becoming more common in the datacenter class of CPUs coming to market in response to the needs of special applications, like those found in the 5G vRAN.

From a historical perspective this advancement should be familiar; the current fixed function ASIC or more programmable eASICs will in time yield to more general-purpose hardware. The key here is going to be the in-time aspect. As has been mentioned several times now, the CSP community as a whole are very concerned with the TCO of the solutions that go into their network and again, new technology when initially introduced can struggle to compete with the legacy technology it seeks to replace. But just as the "unreliable and noisy" transistor eventually displaced all but the boutique vacuum tube, it is reasonable to expect that more and more of the RAN workload will find its way into the CPU. One key element as a minimum is the enablement; that is, once the CPU with the required features becomes available, along with a compiler that can recognize the FP16 data structures and generate the low-layer code to use these special registers, the development community will also have to skill up to transform the lower-layer functionality from hardware into the language of the CPU. The careful reader may notice that we have skipped right over the transition from ASIC and eASIC to FPGA and then to the CPU realization. While this was not an oversight on our part, today we see the use of the FPGA as a short-lived transition from fixed function ASIC to a general-purpose compute solution, provided the CPU can efficiently realize the workload.

One final thought on FP16: the range of numbers that can be represented is limited, with 1 bit allocated to a sign bit, 8 bits allocated to the exponent, and the remaining 5 bits allocated to the significant portion of the value to be represented, the application in the mobile space where a limited constellation of values has proven useful for conserving the compute resources allocated to the task at hand.

## 7.7 Enhanced Common Public Radio Interface: The Fronthaul Interface Transformation

One of the key efforts to transform the way the network is built requires the ability to separate nodes from a single-source vendor into one where connected nodes across the front-haul interface can be sourced from multiple vendors. The open enhanced Common Public Radio Interface (eCPRI) protocol specification [10] was developed in collaboration between four companies that would otherwise be viewed as competitors in the industry: Ericsson, Huawei, NEC, and Nokia. Figures 7.12 and 7.13 show the separation of the RF portion of the RAN from the DU, and this interface in an open design conforms to the eCPRI protocol.

The fronthaul interface (also called fronthaul transport network) connects eCPRI-capable radio equipment to an eCPRI-capable radio equipment controller. An additional motivation for this specification is to reduce the data rate demands over the interface, and the functional decomposition provides flexibility to limit the complexity of the radio equipment when compared to legacy CPRI. The flexibility found in eCPRI, such as using packet-based transport, IP over Ethernet facilitates the various physical layer splits within the base station design. The three different elements consisting of the user data, the control and management (C&M) data, and the synchronization planes are all transported within the eCPRI protocol. Fundamentally, this specification defines a new layer above the transport layer for the user plane and retains standard protocols for the synchronization as well as for the control and management layers.

The specification also provides support for cases requiring interworking between two radio elements and for interworking to legacy CPRI.

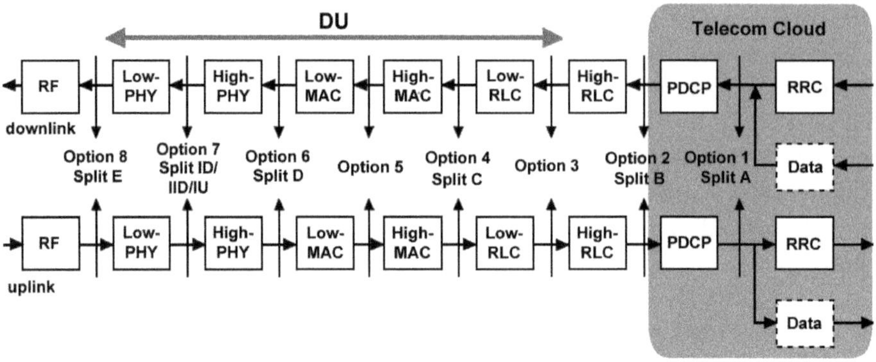


**Figure 7.12** eCPRI split eREC and eRE.

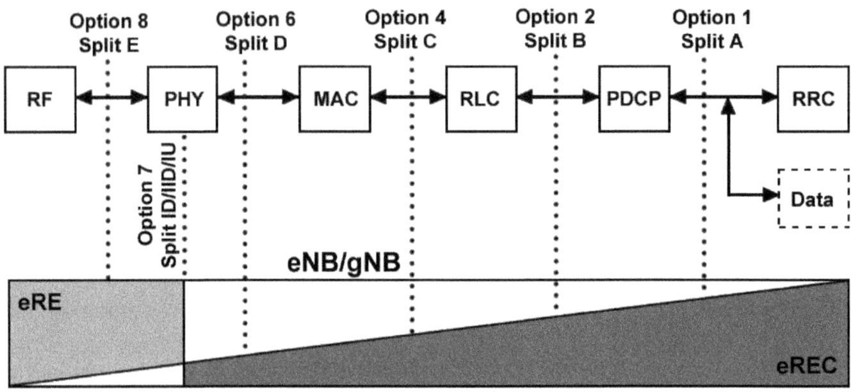


**Figure 7.13** eRE and eREC split options.

Within the specification there is a requirement that a radio base station consists of (at least) two eCPRI nodes. The minimum is one radio element, called the eRC in the specification, and the second is the radio equipment controller or the eREC node.

The synchronization data plane requires the presence of a grandmaster clock (GM). This clock is derived from a Global Positioning System (GPS) sourced clock and is required to maintain network synchronization across multiple base stations, and to allow for the precise control of timing for UE control. This one element of the mobile network is a significant divergence from other protocols found in the wireless domain. The grandmaster clock may be located in the network or in either of the two eCPRI nodes. Note that this clock is often found a node or two deeper in the network for several practical reasons, and the replication of the timing can then be distributed to the elements closer to the RF interface. It is common in disaggregated designs for this clock to be part of a NIC. A stratum 1 is often found deep in the CSP network, actually multiple instances for redundancy, and the network elements function on a stratum 2 timing source. The baseline reference is the stratum 0, which is the highest level of reference clock sourced from the Global Navigation Satellite System (GNSS) (most often from the well-known GPS system of satellites) and national time and frequency radio signals. Within the mobile network the Network Time Protocol (NTP) is used to implement the timing references.

The fronthaul network shown in Figure 7.14 exists as the interface between the two nodes. Irrespective of the separation, the eRE is very likely in close proximity to the antennas in every case as this node is responsible for the analog radio frequency functions, and the eREC implements the higher-layer air interface functions.

In the case when the eRE and the eREC are both located at the tower site, the fronthaul is realized using copper or fiber interfaces, whereas when the eREC is distant from the tower location the choice of the physical medium is nearly fiber in every case. While there may be some localities that still use copper, these are rapidly diminishing. There are use cases where the fronthaul is realized by direct connection to a standard high-volume server at each node. In this case there may not be a long-haul capable cell site router located at the base of the tower and as such, the compute nodes at both the eRE and eREC must be capable of driving the optical power necessary to span the distance. Not all small formfactor pluggable (SFP) fiber interface modules are created equal. The use of long-haul SFPs is necessary and recent deployments have revealed that not all NIC card drivers may support the selected and necessary SFPs needed for

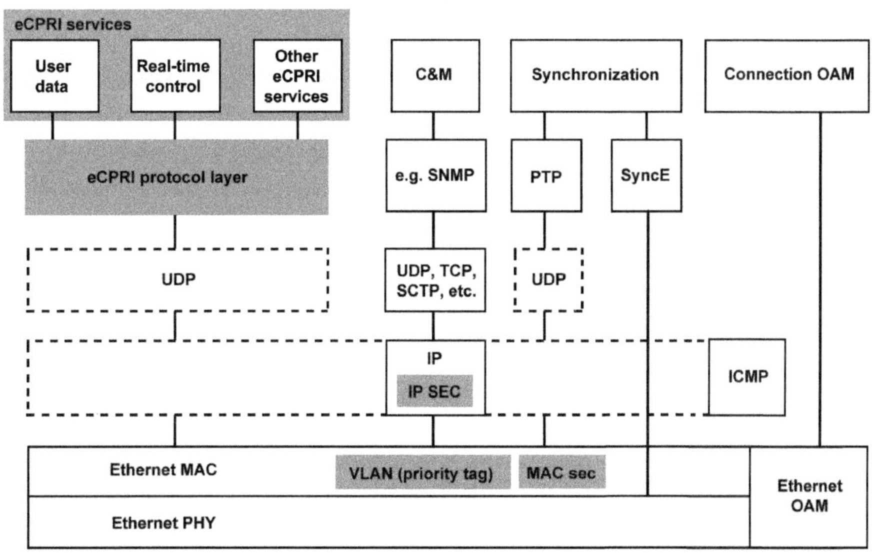


**Figure 7.14** eCPRI protocol stack.

the long-haul direct connection between the nodes. This may be derivative of the use of standard high-volume servers that are very common in datacenter applications where the server to top of rack (TOR) or end of rack (EOR) distances are relatively short. The long-haul SFPs have a higher power requirement, nearly twice that of the SFPs used in shorter datacenter applications. While twice the power may sound ominous, in reality the specification increases the demand on the server by about 1.5W per SFP. As a result, a 1,200W server (based on the power supply capability), operating at around 500W for a dual-socket system with two NICs requires about 3W for the datacenter-capable SFPs and around 6W for the long-haul-capable SFPs. This increase in the power budget may be insignificant when compared to the normal variation of the heat load (power consumed and dissipated) due to other environmental conditions of operation. Nevertheless, these real-world conditions continue to be discovered when the standards like eCPRI meet the real-world deployments.

The implementation of the physical layer split as spelled out in the eCRPI standard is no small task and can be found in several open-source projects. The current implementations may rely on the capability of lookaside accelerators until such time as CPUs with native support for some of the functions become available as discussed earlier.

## 7.8 Summary

Virtualization of the RAN remains a goal. While some progress has been made, there are still serious concerns over the power and performance. Eventual growth into the mmWave from the current sub-6-GHz implementations will require significant work. Small cell deployments where venues with transient high concentrations of subscribers, such as those found at sporting and concert venues, will continue to offer innovation opportunities but at the same time will be challenging on the revenue generation front. The recent introduction of private wireless networking, which may appear under a multitude of other labels such as private networks and p5Gs, are starting to see limited deployment in people-scarce, resource-dense enterprises, some indoors, some outdoors. There are several constraints here; one is the availability of spectrum. In many jurisdictions the controlling regulatory authorities have not made spectrum available, and enterprises remain dependent on obtaining spectrum from the carriers. This can complicate or stall introduction and innovation. In other jurisdictions, such as the United States, some lightly licensed spectrum or some other similar scheme has been implemented to open up spectrum for these use cases. The CBRS spectrum in the United States is shared and a system has been put in place to manage the allocation of the spectrum. Here the opportunity for virtualized RAN to be deployed may gain a significant market foothold, allowing vendors to further mature their offerings (including reducing power demands among other goals) while the carriers continue to transform their networks to accommodate more virtualized solutions [11].

## References

[1] https://www.rfc-editor.org/rfc/rfc8259.

[2] https://www.rfc-editor.org/rfc/rfc4960.

[3] https://www.3gpp.org/release-15 3GPP Release 15.

[4] https://www.3gpp.org/release-16 3GPP Release 16.

[5] https://www.3gpp.org/release-17 3GPP Release 17.

[6] https://www.etsi.org/deliver/etsi_tr/121900_121999/121905/06.10.00_60/tr_121905v061000p.pdf.

[7] https://www.etsi.org/deliver/etsi_ts/129200_129299/129281/15.03.00_60/ts_129281v150300p.pdf.

[8] https://www.resurchify.com/5G-tutorial/5G-NR-PDCP.php.

[9] https://standards.ieee.org/ieee/754/6210/.

[10] http://www.cpri.info/downloads/eCPRI_v_2.0_2019_05_10c.pdf.

[11] https://www.intel.com/content/www/us/en/docs/programmable/683685/22-1-1-4-1/ecpri-ip-source-interface.html.

# 8

# vRAN Performance Engineering

## 8.1 Network Performance Engineering

Many network operators and network equipment providers have 4G functionality running within the same software and on the same hardware as 5G elements. This decision is important for several technical and business reasons. In this chapter, we will investigate the requirements of having standalone and non-standalone functionality in the network, or, as we have previously termed it, how the metal is stretched around the specifications.

### 8.1.1 5G Drivers

While previous generations of mobile networks are purpose-built for delivering specific and limited communications services such as voice and messaging (e.g., 2G) or mobile broadband (e.g., 4G), 5G must have flexibility and configurability at the heart of its design to enable the introduction, over time, of new use cases in support of smart cities, smart agriculture, logistics, public safety agencies, and so on.

### 8.1.2 5G Usage Scenarios

5G refers to mobile systems that fulfill the ITU-2020 requirements [1]. The ITU M.2083-0 report defines the framework and overall objectives of the future development of international mobile telecommunications (IMT) for 2020 and beyond to better serve the needs of a networked society. IMT-2020 is expected to expand and support diverse use cases and applications that will continue beyond the current IMT framework. These use cases are broadly seen as a continuum of options with varying requirements as follows (see Figure 8.1).

eMBB is conceived to address the human-centric use cases for access to multimedia content, services, and data. Typical applications include three-dimensional (3-D) video, ultrahigh-definition (UHD) screens, and augmented reality.

URLLC is characterized by stringent requirements for mobile network capabilities such as throughput, latency, and availability. Typical applications include industry automation, mission-critical services, and self-driving cars.

mMTC encompasses scenarios that comprise very large numbers of connected devices that individually transmit a relatively low volume of delay-tolerant data. Typical applications include smart grid, smart home/building, and smart cities.

By supporting these generally defined requirements, 5G systems have the intrinsic flexibility to service a variety of use cases including those that are unforeseen at this time and are yet to be identified. Figure 8.1 shows only a few examples of usage scenarios that are envisioned for IMT by 2020 and beyond.

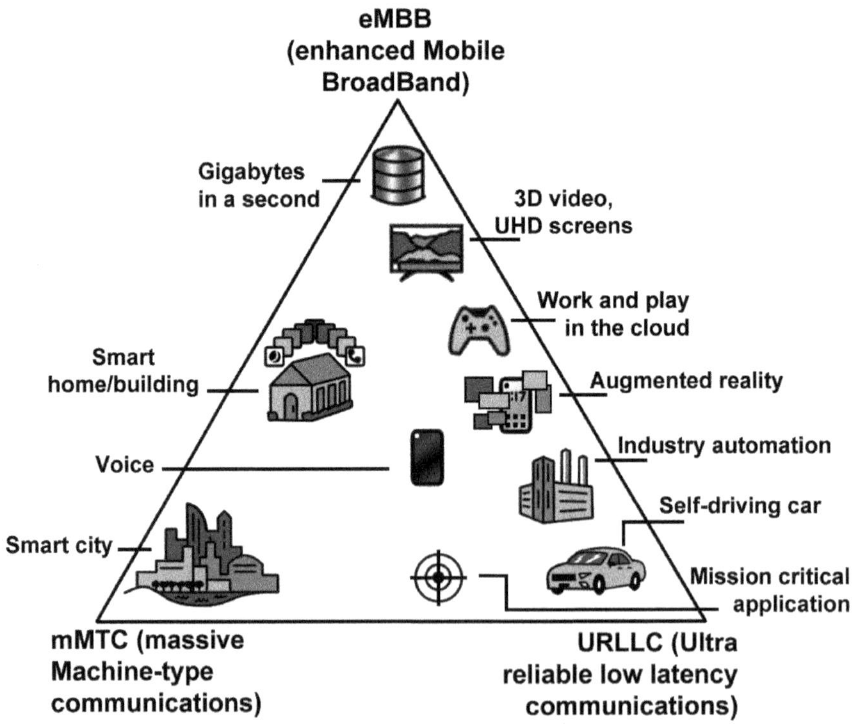


**Figure 8.1** Framework and use cases of IMT for 2020 and beyond [2].

### 8.1.3 5G Spectrum Bands

The spectrum bands earmarked for 5G deployment can be subdivided into three macrocategories: sub-1 GHz, 1–6 GHz, and above 6 GHz.

- Sub-1-GHz bands are suitable to support IoT services and extend mobile broadband coverage from urban to suburban and rural areas. This is because the propagation properties of the signal at these frequencies enable 5G to create very large coverage areas and deep in-building penetration.
- The 1–6-GHz bands offer a reasonable mixture of coverage and capacity for 5G services. A reasonable amount of existing mobile broadband spectrum resources has been identified with this range, which is expected to support initial 5G deployments.
- Spectrum bands above 6 GHz provide significant capacity (thanks to the very large bandwidth that can be allocated to mobile communications) and thus are highly suitable to enable eMBB subject to reduced signal coverage and limited ability to penetrate structures.

With these macrocategories of spectrum band allocation, current and future 5G deployments should offer the versatility needed to support a wide variety of use cases and their respective performance requirements. In addition, operators have the opportunity to evaluate the feasibility of different technologies to best support their use cases of interest. The global interoperability feature will help operators ensure that their networks deliver their use cases effectively.

## 8.2 5G Functional Split

### 8.2.1 5G Functional Split Origin

The centralized RAN (C-RAN) concept was introduced a few years ago to take advantage of the centralization of baseband unit (BBU) functions and achieve improved radio and computing resource allocation, energy efficiency, cost-saving, and so on. In addition to resource efficiency, C-RAN brings agility, flexibility, and time-to-market acceleration of future network features, which can be implemented via software upgrades without requiring hardware upgrades. In its Release 15, 3GPP defines a new, flexible architecture for the 5G RAN, where the base station—or gNodeB (gNB)—is functionally split into three logical nodes: the CU, the DU, and the RU as shown in Figure 8.2. Each unit so defined is capable of hosting specific and distinct functions of the 5G New Radio (NR) protocol stack. Possible *functional split options* are discussed in Section

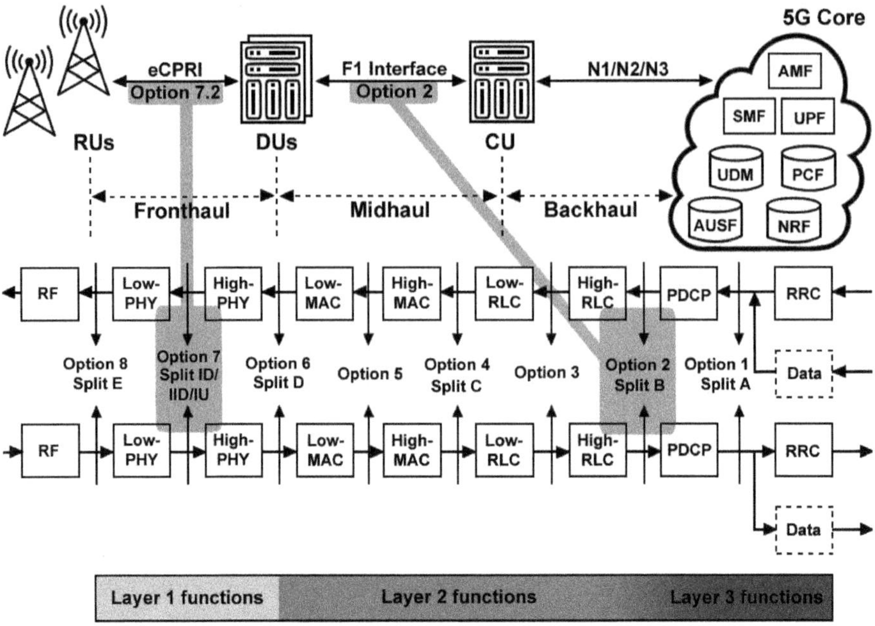


**Figure 8.2** 3GPP functional splits.

8.2.3. Synergistically to these 3GPP's definitions, a number of other standardization bodies (IETF, ITU) [3, 4] have been focusing on the definition of the new transport network interfaces: fronthaul, mid-haul, and backhaul, as shown in Figure 8.2. The fronthaul network is a key enabler to achieving flexible RAN deployments, while the latter two types are significantly dependent on the chosen functional split option.

### 8.2.2 eCPRI

The Common Public Radio Interface (CPRI) is a serial interface that is designed to send constant bit rate (CBR) data from the remote radio transceivers—referred to as either remote radio units (RRUs) or remote radio heads (RRHs)—to the BBU. The CPRI link transports digitized RF signals in a complex baseband format. The eCPRI standard has emerged as a successor to CPRI. Developed by the eCPRI Forum, this protocol makes more efficient use of the transport network bandwidth compared to its predecessor and is *packet-oriented*, implying that it can be framed within the Ethernet protocol. The packet-oriented nature of eCPRI brings tangible advantages to the fronthaul network design which, depending on the functional split, can now make use of Ethernet connectivity instead of relying on the availability of point-to-point fiber-optic cabling between the RU and DU. Being defined as an open interface, eCPRI enables operators to mix and match vendor equipment.

### 8.2.3 Functional Split Options

When describing the functional split options, both the CU and DU may be considered as a single logical unit because the splits apply to the fronthaul network which connects the RU to the DU. However, higher-level functions of the protocol stack may be distributed between the CU and DU that communicate over the mid-haul network. Bit rates and latency requirements over the mid-haul transport network are determined by the higher-layer stack functions. In practice, the industry has settled on a split between CU and DU where CU hosts the network layer Radio Resource Control (RRC) functionality and the PDCP functionality.

Figure 8.2 shows the eight commonly used splits, with each option offering a distinct trade-off between centralization benefits and fronthaul network requirements. For example, Option 8 corresponds to the original C-RAN configuration in which all stack functionalities are centralized. Option 8 is defined to maximize the benefits of adopting a fully centralized baseband processing architecture that enables load-balancing and sharing of the processing capability across the deployed RUs.

### 8.2.4 Functional Splits Trade-Off

An additional advantage of centralizing the baseband functionality is the possibility of implementing many of the network functions virtually, with the software hosted on commercial-off-the-shelf (COTS) servers. Eventually, the only radio unit functions that must be implemented in the remote site are the RF elements, making the RU remote site small and easy to host on the tower, inexpensive, and with reduced power consumption. Operators can perform most network upgrades at the CU, thus requiring fewer visits to the remote sites in the field. The simplified RU can handle multiple radio access technologies (RAT), further reducing the footprint of the remote site that is expected to support multiple cellular generations.

By fully centralizing the functionality of the 5G RAN stack, Option 8 places the most stringent demands on the fronthaul network resulting in high transmission bit rates and strict latency requirements. In fact, the information flow at the lowest PHY layer is in its most raw format, essentially requiring radio signals to be sampled and encoded into binary values before being transported over the fronthaul network in both DL and UL directions. Functions in the PHY layer such as the handling of cyclic redundancy checking (CRC) bits, signal modulation, mapping, and encoding, add more information to the data blocks handled at the higher MAC layer. Thus, progressively higher bit rates occur as the information flows toward the PHY RF function (left in Figure 8.2) and vice versa, lower bit rates occur as the information flows toward the upper layers (right in Figure 8.2).

Additionally, the time-sensitive nature of the data exchange that takes place between the PHY layer and certain higher-level processes such as the hybrid automatic retransmission request, (HARQ) implemented at the Medium Access Control (MAC) layer requires network round-trip delays as low as 5 ms. The resulting fronthaul requirements of Option 8 effectively limit its applicability to scenarios where fiber-optic infrastructure is economically justifiable, such as in urban areas or where operators already own it.

At the opposite end of the functional split options, the Option 1 split places all the baseband processing functionalities within the RU, which is consequently physically large, complex, and requires more power compared to the simplified RU of Option 8. The fronthaul network requirements of Option 1 are, however, significantly relaxed because the entire protocol stack resides in the RU. In fact, more processing occurs before data is transported from RU to CU, leading to much lower bit rates and higher latency tolerance for the fronthaul network. The remaining functional split options, from 2 to 7, facilitate a range of solutions enabling varying trade-offs between the amount of baseband processing required in the RU (as opposed to being hosted in the DU/CU) and the fronthaul network capabilities, as illustrated in Figure 8.2.

### 8.2.5 How to Select and Additional Functional Split Options

The choice of RAN split in 5G deployment depends on the specific radio network deployment scenario, intended supported services, existing infrastructure, and cost targets. Additional influencing factors include:

- Specific QoS requirements pertaining to the offered services (e.g., low latency, high throughput);
- Network densification, traffic profile, bandwidth demand, available spectrum, and advanced RF features like multiple-input multiple-output (MIMO) antennas;
- E2E transport availability networks with different performance levels, from ideal to nonideal;
- Specific use cases: real-time or non-real time.

Choosing the RAN split also depends on end-to-end related aspects of mobile networking that include transport, data center placement and connectivity, packet core, infrastructure programmability, automation, data collection, and analytics.

#### 8.2.5.1 Key Split Options for Initial Deployment

Initial 5G deployments may favor certain split options over others due to availability of technology. The CU-DU split design with distributed capability and virtual functionalities defined by the 3GPP 5G RAN specification is further enhanced by new variants and flexible topologies that are being proposed by both the O-RAN Alliance [5] and the Small Cell Forum [6] as described next.

*Split 2: RRC/PDCP split.* RRC and PDCP are split from the Layer 2 radio link control (RLC). Two flavors are available for this Split 2 option:

- Variant 1: RRC, Service Data Adaptation Protocol (SDAP; represented as Data in Figure 8.2) and PDCP are implemented in the same CU as one entity; no control and user plane split. RLC, MAC, and High PHY layer are implemented in the DU, while the Low PHY layer along with the RF is implemented in the RU.
- Variant 2: In addition to Variant 1, this split option comes with the further separation of the control and user plane to be implemented in the gNB-CU-Control Plane (CU-CP) and gNB-CU-User Plane (CU-UP), respectively. More specifically, RRC and PDCP-C (the control plane part of PDCP) are implemented in the CU-CP, while SDAP/PDCP-U (the user plane part of PDCP) are implemented in the CU-UP.

*Split 6: MAC/PHY layer split.* The MAC, RLC, and upper layers are implemented in the CU. The full stack of the PHY layer (Low PHY and High PHY) and the RF are in the DU/RU.

*Split 7.2x: Low PHY/High PHY split.* The Low PHY/High PHY split is the most acceptable approach for it is less complex and supports various fronthaul requirements, and most important, it has high virtualization benefits. This split has been further optimized by the O-RAN Alliance into two variants: split 7.2a and split 7.2b. Split 7.2x comes with fronthaul compression techniques like block floating point (BFP) of the in-quadrature (IQ) signal representations to further reduce transport data rate requirements.

*Split 8: PHY/RF split.* This 3GPP defined split was initially considered in traditional C-RAN based designs that make use of CPRI in support of RRUs. It has been determined that this split also holds benefits in some unique business cases of 5G. This split option enables complete separation of the RF from the PHY layer to maximize virtualization gains. Everything from the physical layer and up, including all protocol layer levels, is centralized, resulting in a very tightly coordinated RAN. This option allows efficient support of advanced 5G features that require extreme low latency like multitransmission and reception point (TRP) [7], high-order MIMO, and high diversity for URLLC-like traffic.

## 8.3 5G Deployment Options: SA and NSA Architecture

Previous wireless generations require that both access and core network be of the same generation (e.g., EPC and LTE must work together to realize a 4G system). Departing from this requirement, 5G is designed to allow integration of elements from different generations using two configurations: standalone and non-standalone.

3GPP defines both a new 5G core network, referred to as 5GC, and a new radio access technology, 5G NR. The 5GC is designed to be cloud-native, thus inheriting many of the technological solutions used in cloud computing and with virtualization at its core. The NG-RAN node (i.e., base station) in the 5G is either a gNB (i.e., a NR base station), providing NR UP and CP or a ng-eNB (i.e., an evolved LTE base station), providing LTE services to the legacy user equipment. Refer to Figure 8.3 for a pictorial representation of these options.

In standalone scenarios, either the 5G NR (Option 2) or the evolved LTE radio cells (Option 5) are operated independently with the 5G core network. This means that the NR or evolved LTE radio cells are used to support both

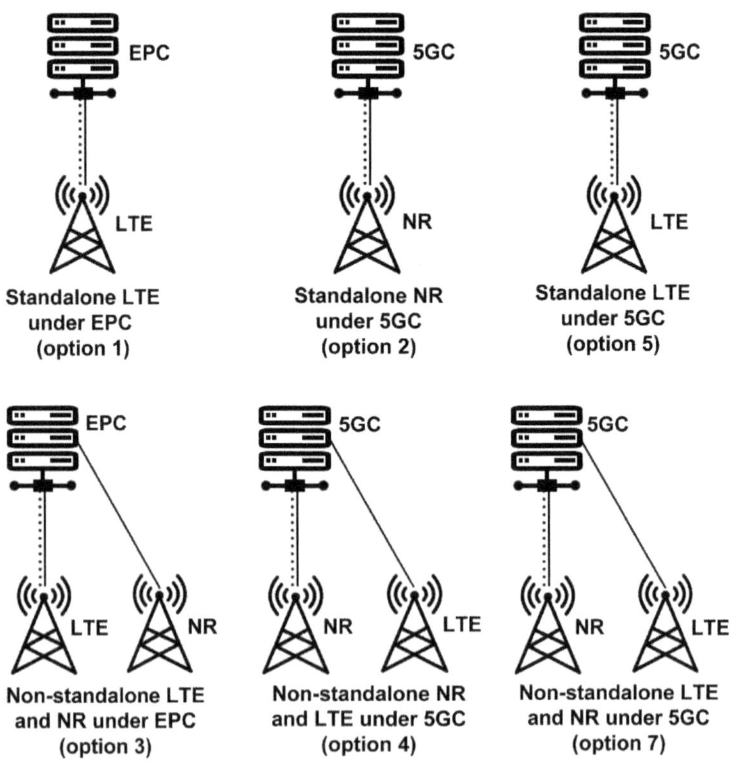


**Figure 8.3** SA and NSA deployment options.

user plane (solid line) and control plane (dotted line). The standalone option is a simple solution for operators to manage and may be deployed as an independent network using normal intergeneration handover between 4G and 5G for service continuity. Each standalone architecture uses only one radio access technology.

In NSA scenarios (Options 3, 4, and 7), the NR radio and LTE radio cells are attached to the same core network (either EPC or 5GC). NSA enables operators to leverage already existing 4G deployments, connecting LTE and NR radio resources to existing EPC and/or newly deployed 5GC resources.

### 8.3.1 SA and NSA Deployment Options

5G deployment options are being defined in 3GPP using either the existing EPC (which is specified in 3GPP TS 23.401) or the 5GC (which is specified in 3GPP TS 23.501).

Two variations of SA are being defined in 3GPP:

- Option 2: using 5GC and NR gNB access;
- Option 5: using 5GC and LTE ng-eNB access.

In the SA Option 2, gNBs are connected to 5GC through the NG interface [8]. The gNBs communicate with one another through the Xn interface. In the SA Option 5, ng-eNBs are connected to 5GC through the NG interface. The ng-eNBs communicate with one another through the Xn interface. Essentially, Option 5 allows the existing LTE radio infrastructure (through an upgrade to the eNB) to connect to the newly deployed 5G Core.

Three variations of NSA are defined in 3GPP:

- Option 3: using EPC and a LTE eNB acting as master and NR en-gNB acting as secondary;
- Option 4: using 5GC and a NR gNB acting as master and LTE ng-eNB acting as secondary;
- Option 7: using 5GC and a LTE ng-eNB acting as master and a NR gNB acting as secondary.

In the NSA Option 3—commonly known as Multi-Radio Access Technology (Multi-RAT), LTE-NR Dual Connectivity (EN-DC)—a UE is connected to an eNB that acts as master node (MN) and to an en-gNB that acts as secondary node (SN). An en-gNB is different from a gNB in that it only implements part of the 5G base station functionality, which is required to perform SN functions for EN-DC. The eNB is connected to the EPC via the S1 interface and to the en-gNB via the X2 interface. The en-gNB may be also connected to the EPC via the S1-U interface and to other en-gNBs via the X2-U interface. Notice that the en-gNB may send user-plane packets to the EPC either directly or via the eNB (through secondary bearer split).

In the NSA Option 4, a UE is connected to a gNB that acts as a MN and to an ng-eNB that acts as a SN. This option requires the 5G Core to be deployed. The gNB is connected to 5GC and the ng-eNB is connected to the gNB via the Xn interface. The ng-eNB may send user-plane packets to the 5G Core either directly (Option 4a) or through the gNB (Option 4).

In the NSA Option 7, a UE is connected to a ng-eNB that acts as a MN and to a gNB that acts as a SN. The ng-eNB is connected to the 5GC, and the gNB is connected to the ng-eNB via the Xn interface. The gNB may send user-plane packets to the 5GC either directly or through the ng-eNB.

### 8.3.2 Technical and Cost Comparison

#### 8.3.2.1 Technical Comparison between SA and NSA Options

A technical comparison between the 5G SA and NSA options for mobile network is summarized in Table 8.1. For simplicity, only the NSA Options 3, 7, and the SA Option 2 are reported.

**Table 8.1**
Technical Comparison between SA and NSA Options

| | **NSA** | | |
|---|---|---|---|
| | **Option 3** | **Option 7** | **SA (Option 2)** |
| **3GPP 5G Specification** | Rel-15 (17.12) | Rel-15 (18.6) | Rel-15 (18.6) |
| **5G Spectrum** | Sub-6 GHz and mmWave | Sub-6 GHz and mmWave | Sub-6GHz |
| **Core Network** | EPC | 5GC | 5GC |
| **Core Network Interworking** | Not required | Not required | Required between 5GC and EPC |
| **UE Connectivity** | EPC-NAS | 5GC/EPC-NAS | 5GC/EPC-NAS |
| **Network Slicing** | Not supported | Supported | Supported |
| **5G QoS** | Not supported | Supported | Supported |
| **3GPP Release Forward Compatibility** | Low | Mid | High |
| **RAN Interworking** | EN-DC | NGEN-DC | NR-DC |
| **LTE Upgrade** | Required for eNB and EPC | Required for ng-eNB and 5GC | None or minor |

*Spectrum:* Due to its coverage limitation and cost burden, mmWave band cells are only recommended in SA mobile networks, while sub-6-GHz band is desirable in all shown options.

*Core network:* Noting that Option 7 and Option 2 require a new 5GC system, these options first require the development and deployment of 5GC. Conversely, enhancement of EPC (such as control and user plane separation of evolved packet core or CUPS) for Option 3 is desirable to accommodate large NR capacity. Interconnection interfaces between 5GC and EPC are only required in a SA network to support intersystem mobility. When it comes to 5G-specific services, 5GC can naturally support them through the new QoS framework and network slicing features. Option 3 requires a single EPC Non-Access Stratum (EPC-NAS) protocol to manage the establishment of communication sessions with the UE as it moves, keeping the UE complexity lower than that of Option 7 or SA Option 2.

*RAN:* Because the NSA networks can fully leverage existing LTE infrastructure coverage, it facilitates quick deployment of a full-coverage 5G network with relatively small investment. For Option 3, deployment requires an upgrade of LTE eNB and EPC based on EN-DC. For Option 7, a significant upgrade of LTE ng-eNB to support NGEN-DC dual connectivity is required, along with the introduction of 5GC.

#### 8.3.2.2 Deployment Time and Cost Comparison between NSA and SA Options

The incurred cost, deployment time, and service support that are expected for the 5G NSA and SA options for mobile network are summarized in Table 8.2. Note that cost and time estimations may vary as they heavily depend on specific technical and business parameters.

**Table 8.2**
Deployment Time and Cost Considerations between 5G SA and NSA Options

| | **NSA** | | |
|---|---|---|---|
| | **Op. 3** | **Op. 7** | **SA (Option 2)** |
| **Required Time for Deployment** | Short | Medium | Long |
| **Deployment Cost** | Low | Mid | High |
| **Cost for System Upgrade** | Low | High | Low |
| **Migration Cost (NSA to SA)** | High | Medium | None |
| **New 5G Services Support** | Not supported | Supported | Supported |
| **Deployment Service Quality** | Medium | High | Low |
| **NR Coverage Quality** | Low | Low | High |
| **Voice Service for 5G UE** | CSFB and VoLTE | CSFB and VoLTE | VoLTE and VoNR |

*Deployment and cost:* Since NR cells can be deployed in urban hotspot areas as capacity and speed boosting cells in NSA network, deployment time will be shortened and NR CapEx will be lower than that of an SA network. However, more investment will be required to upgrade existing LTE RAN for Option 7. In addition, 5G spectrum acquisition cost for SA is higher than that for NSA due to the sub-6-GHz spectrum that is required for nationwide coverage in the former option.

*Migration:* The initial deployment of a NSA network causes additional cost and deployment time when the time comes to migrate to a mature SA network. The LTE upgrade costs are high for Option 7 due to the nature of the upgrades and the introduction of 5GC.

*Services:* Option 3 offers limited support for 5G-specific services due to its use of legacy EPCs. On the other hand, SA Option 2 natively provides best NR coverage quality while it is expected to initially offer limited nationwide coverage.

### 8.3.3 Migration Path from 4G LTE to 5G

One of the highly recommended paths for vendors and operators is migration from Option 1 → Option 3 NSA → Option 7 NSA → Option 4 NSA or Option 2 SA. This migration path is optimal for operators that have a widely deployed 4G network and aim to smoothly evolve to SA 5G and develop extended 5G/6G services. With this migration path operators can deploy 5G with the lowest CAPEX investment and progress toward intermediate steps over a prolonged period.

#### 8.3.3.1 Deployment Considerations for Option 1 to Option 3 Migration

Depending on the EPC features defined by 3GPP in Release 15 and future releases, the EPC capabilities may represent a possible bottleneck (e.g., latency), which limits the performance that could otherwise be extracted from NR. Data throughput per 5G connected subscriber (e.g., in SGW/PGW) is expected to increase via NR and LTE in dual LTE-5G connectivity. From the end-user device point of view, the attractiveness of this solution is that it only requires the additional support for the NR specifications. The device will communicate with the core network using the same EPC procedures used by currently available devices, either under only LTE or under both LTE and NR coverage. However, it should be noted that combining LTE and NR radio interfaces for split bearers may increase the device memory requirements.

#### 8.3.3.2 Deployment Considerations for Option 3 to Option 7 Migration

This migration requires UEs to support the new protocol stack to access the 5G core network (5GC). Already in the field, LTE RANs must be upgraded to connect to 5GC, and more LTE base stations (eNodeB) need to be upgraded to interwork with NR. This migration path also requires tight interworking between LTE (already widely deployed) and NR (operators can selectively deploy NR only where needed).

#### 8.3.3.3 Deployment Considerations for Option 7 to Option 4 Migration

Option 4 is applicable in areas where NR provides wide coverage and is overlaid by LTE, enabling use of NR as the master technology. Option 4, Option 5, and Option 7 are complementary to standalone Option 2 in supporting a faster adoption of the 5GC in a wider variety of deployments. Noting that Option 4 is the only one that makes NR the master technology, more network areas become suitable for Option 4 deployments as NR spreads widely. Option 4 removes the inherent NSA evolution constraint of Option 7 whereby NR is dependent on LTE to provide the anchor connection.

#### 8.3.3.4 Deployment Considerations for Option 7 to Option 2 Migration

Option 2 provides full support for new 5G applications and services including eMBB, mMTC, and URLCC. The coverage is totally based on the NR bands. Like Option 4, in Option 2, NR acts as the master node except for the support of multi-RAT dual connectivity.

Other considered migration paths include:

1. Option 1 → Option 7 → Option 4/2: Even though operators will incur higher investment costs, this path is highly suitable for operators that already have NFV, SDN, and 4G coverage. This migration path utilizes 4G RAN infrastructure and spectrum to launch 5G services. This migration path is useful for operators who have already high traffic on their large deployments and high-frequency spectrum.
2. Option 1 → Option 3 → Option 4/2: With this migration path, the operators can deploy early 5G with minimal investment cost. Operators with rapid growth of 5G traffic can leverage their existing 4G and directly expand into 5G coverage.
3. Option 1 → Option 4/2: This option is recommended for special use cases such as delay-sensitive applications. NR works in SA mode and 5GC is the chosen CN. Facing initial high investment costs, this migration path requires forward compatibility with 3GPP Rel.16/17 and is suitable to operators who have spectrum availability for the wider coverage of NR in SA mode.

## 8.4 5G Roadmap

The 3GPP technical specification effort is a global system engineering project. Participants (including typically vendors and operators) participate in the creation of specifications from initial R&D through final product, conforming to the specification.

### 8.4.1 3GPP Release of 5G NR

The 3GPP Release 15 (completed in 2018) and 16 (completed in 2020) specifications, also known as 5G Phase 1 and Phase 2, respectively, focus mainly on the URLLC. Release 17 mainly focuses on the 5G enhancements.

The major characteristics of 5G NR, Release 15 (i.e., Phase 1) that enables eMBB (10–20 Gbps) include:

- Ultrawide bandwidth (up to 100 MHz in <6 GHz, up to 400 MHz in >6 GHz);
- Set of different numerologies for optimal operation in different frequency ranges;
- Native forward-compatibility mechanisms;
- New channel coding;
- Native support for URLLC;
- Flexible and modular RAN architecture: split fronthaul, split control and user plane;
- Native end-to-end support for network slicing.

The major focus in Release 16 (i.e., Phase 2) is to enable URLLC for mission-critical applications. The major characteristics include:

- Enhancement of URLLC;
- Cellular IoT support and evolution;
- Advanced V2X support;
- 5G Location and positioning services;
- UE radio capability signaling optimization;
- Enablers for network automation architecture for 5G;
- Wireless and wireline convergence enhancement;
- Mission-critical, public warning, railways, and maritime;
- Streaming and TV;

- User identities, authentication, multidevice (network) slicing.

The building blocks of 5G that enable URLLC include:
Low latency: NR

- Enables shorter slots in the radio subframe.
- Introduces a mini-slot; prioritized transmissions are started without waiting for slot boundaries.

Time synchronization

- The radio network components are time synchronized, for instance through the ITU Precision Time Protocol (PTP) telecom profile with full timing support for the network.

Resource management: NR introduces

- Preemption, where URLLC data transmission can preempt ongoing non-URLLC transmissions.
- Fast processing, enabling retransmissions even within short latency bounds.

Reliability

- Via extra-robust transmission modes for both data and control radio channels.

### 8.4.2 5G Services in North America

At the time of writing, North America is currently one of the most developed 5G markets globally with mobile 5G subscriptions at 20.9 million for year-end 2020, equivalent to 3.93% of total mobile subscriptions in the market. It is estimated that by 2023, up to 32% of North America mobile connections will be on a 5G network [9].

#### 8.4.2.1 5G Networks in the United States

Verizon's 5G deployments are focused on mmWave spectrum reaching 4-Gbps peak speeds in some locations. Verizon's 5G Ultra-Wideband service is available in 55 cities, 43 stadiums/arenas, and seven airports.

AT&T's nationwide 5G coverage uses the sub-6-GHz band, offering speeds slightly better than 4G. They are expanding to several US cities using their 5G+ network using mmWave spectrum with significantly higher data rates.

T-Mobile offers a nationwide 5G network using sub-6-GHz bands, which has less capacity than mmWave networks. Following their merger with Sprint, T-Mobile is also deploying its 5G network using Sprint's 5G spectrum. T-Mobile is adding mmWave nodes in a number of U.S. cities.

A survey of the three major U.S. carriers' networks indicates that Verizon's mmWave delivered a maximum download speed of 1.1 Gbps in Chicago. AT&T's 5G+ mmWave network delivered a maximum download speed of 669.2 Mbps in Dallas. T-Mobile, using sub-6-GHz spectrum, delivered a maximum download speed of 213.1 Mbps in Atlanta and over 200.0 Mbps in Chicago and Dallas.

#### 8.4.2.2 5G Networks in Canada

Canadian carriers are aggressively expanding their 5G services.

In early 2020, Rogers Wireless went live with its first 5G NR network in downtown Vancouver, Toronto, Ottawa, and Montreal. Their 5G network rapidly expanded to 130 towns and cities across Canada. Rogers' 5G networks use the sub-6-GHz band and the low-band 600-MHz spectrum. On the 600-MHz bands, Rogers uses Dynamic Spectrum Sharing (DSS) technology to push 4G and 5G simultaneously.

Bell Canada launched commercial 5G service in June 2020 in the metro markets of Montreal, Toronto, Calgary, Edmonton, and Vancouver. They subsequently expanded 5G services to several other markets across Canada. Like Rogers, Bell Canada's 5G services use sub-6-GHz spectrum and leverage existing 4G/LTE cell towers equipped with Ericsson's 5G technology.

Telus Mobility started with the metro markets of Greater Toronto area, Calgary, Edmonton, Montreal, and Vancouver. They subsequently deployed 5G services to communities in British Columbia, Alberta, Manitoba, Ontario, and Quebec.

Even though Canadian operators are mostly using the sub-6-GHz band to offer 5G, a survey in Montreal of Rogers, Bell, and Telus Mobility 5G services recorded consistent performance in handling traffic and speed. A peak download speed of 855 Mbps was achieved on Bell's 5G network, while Rogers excelled in offering more modest but consistent data rates across their network.

### 8.4.3 4G-5G Interworking Architecture

The majority of 5G deployments are brown field with upgrades from existing 3G/4G networks. For mobile providers deploying 5G as green-field new network, it is still necessary to interwork with 4G when roaming under other providers in the same country or internationally.

Figure 8.4 provides a high-level overview of 4G and 5G networks. For clarity, all 4G network functions are in the left column and 5G network functions are in the right column.

The biggest change for any new telecommunications technology is in the access network, also known as the "edge," and 5G is no exception. NR provides new capabilities and allows subscribers to access 5G services. There must be coexistence with legacy 3G and 4G radio at the edge, requiring design and deployment considerations to avoid radio interference, and a handover mechanism enabling users to access the available services based on their device capabilities and subscriptions.

5G NSA architecture enables 5G NR sites to connect to existing 4G Core (4GC). 5G SA provides capability where 5G NR can connect only to 5G Core (5GC) described in Section 8.3.

The 5G network architecture separates control and user planes so that they can scale independently. 5G subscribers are terminated in a SMF.

5G provides much higher bandwidth to subscribers along with support for low-latency services by deploying UPFs at the edge. Multiaccess edge computing (MEC) or xEdge architectures are often considered for this reason.

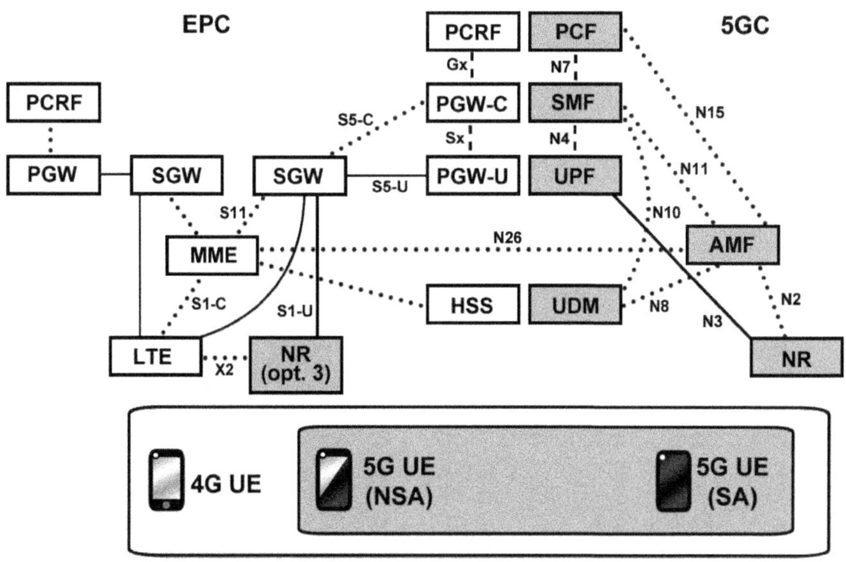


**Figure 8.4** 4G-5G interworking architecture.

### 8.4.4 User Plane and Control Plane Deployment Considerations

Figures 8.5 and 8.6 show the 3GPP 4G and 5G protocol stacks for user and control planes, respectively. The two systems rely on similar architectures and same protocols, except for SDAP. SDAP provides flow-based QoS capability in 5G by providing a mapping between QoS flows and data radio bearers, and marking QoS flow ID (QFI) in both DL and UL packets.

There is a single SDAP entity for each PDU session (GTP Tunnel) (3GPP, 2018e).

In 4G systems, the non-access stratum (NAS) supports mobility management (MM) functionality and user-plane bearer activation, modification, and deactivation. It is also responsible for ciphering and integrity protection of NAS signaling (3GPP, 2018f). In 5G systems, the NAS and MM functionality supports registration management, connection management functionality, and user-plane connection activation and deactivation, as well as encryption and integrity protection of NAS signaling. NAS-session management (SM) is responsible for user-plane PDU session establishment, modification, and release.

As in the earlier 3GPP network releases, the NG-RAN and 5GC have well-defined boundaries, regardless of their implementation. Hence, any security risk in the NG-RAN is managed in the same way as in previous RAN

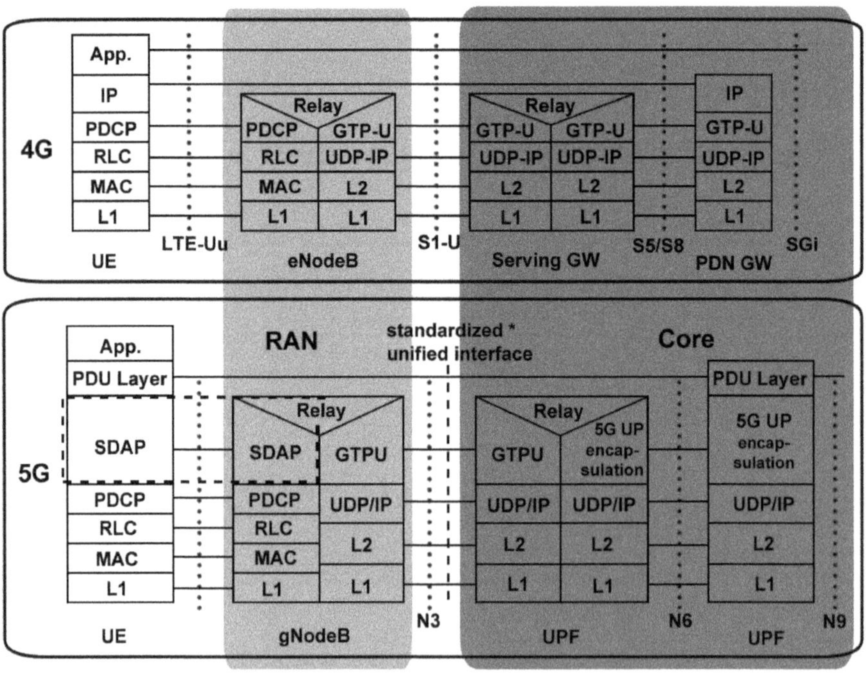


**Figure 8.5** 4G/5G user plane protocol stack.

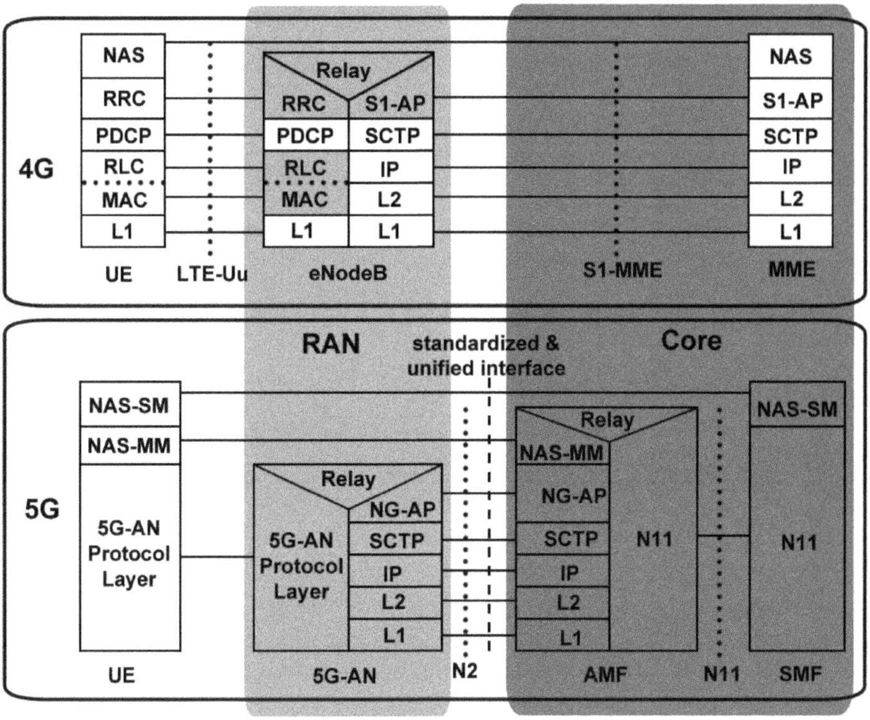


**Figure 8.6** 4G/5G control plane protocol stack

generations. This means that network operators can be selective about the vendor equipment used in the network segments and can pursue a viable multivendor strategy that is cost-effective and mitigates the risk of vendor failure.

## 8.5 Key Challenges in 5G Rollout

5G network designers and engineers face several key challenges that deserve close attention and are therefore discussed in more depth in the following chapters. This last section provides a brief description of two important ones: system security on the one hand, and service performance and availability on the other.

### 8.5.1 System Security

As 5G paves the way to new service opportunities in areas such as healthcare, manufacturing, and transportation, its infrastructure is naturally becoming a more attractive target for cybercriminals due to its increased attack surface and potential value of intrusion. In addition, with 5G there are new and potentially even greater security risks to consider as cloud, data, and IoT threats coexist.

#### 8.5.1.1 Backward Compatibility

It is reasonable to expect that deployed 5G networks will contain 5G security flaws from day one, including those found in existing legacy 4G network core. A security audit on the Diameter protocol, which is used to perform authentication services in 4G, determined that every 4G network contains vulnerabilities that bad actors could exploit to perform illicit actions such as locating users, intercepting SMS messages, and instigating denial of service (DoS) attacks. Consequently, security threats associated with 3G and 4G will persist long after 5G reaches the mass market and will heavily influence 5G operations for a while.

#### 8.5.1.2 Cloud Computing, NFV, and SDN

According to 5G-Public Private Partnership (5G-PPP) [10], 5G will connect 7 trillion wireless devices or "things," reduce the average service creation time from 90 hours to 90 minutes, and enable advanced user-controlled privacy. By connecting most, if not all, aspects of our life, 5G will enable a digital society that expects high service availability and security using a diverse set of technologies. Concepts of cloud computing, SDN, and NFV are combined to meet the growing user and service demands, while at the same time meeting sustainable CapEx and OpEx targets through flexible network operation and management.

Recent studies of these cloud technologies reveal potential security challenges (as well as opportunities) that must be addressed to ensure security of services, infrastructures, and users. For example, multitenant cloud solutions supporting the infrastructures of multiple network operators must provide strict isolation at multiple levels to maintain the integrity of users' information and prevent operators from consuming the resources of other operators, either deliberately or unintentionally.

### 8.5.2 Service Performance and Availability

URLLC relies on features defined by 3GPP 5G-NR Release 15 for mission-critical 5G applications such as the Industrial Internet of Things (IIoT), smart grids, health care, and intelligent transportation systems. URLLC also promises end-to-end security and wireless service up time of 99.999%.

While general-purpose hardware components offer low-cost solutions on the one hand, on the other they are prone to failures that must be overcome through rapid replacement of a faulty component with minimal management effort. Service availability calls for a set of mechanisms to counteract the failure of VNFs and avoid or mitigate the triggered service outages. These mechanisms include reactive and proactive actions upon failures through redundancy mechanisms and VNF migration away from a malfunctioning host.

#### 8.5.2.1 Redundancy Allocation

In cloud-native 5G mobile networks, support for service continuity represents a unique challenge as the software implementing the core network functionalities is decoupled from its underlying hardware. The impact of a node failure and/or VNF failure is high if remedial measures are not implemented by design [11]. Causes of VNF failures may highly vary and depend on the VNF deployment methodology. Regardless of the chosen VNF deployment mode, it is commonly agreed that the following mechanisms are of critical importance: (1) a fast and accurate detection mechanism of failing VNFs, and (2) effective restoration mechanisms that can swiftly provide a VNF replacement. In commercial deployments, it is not uncommon to have a pool of VNFs depending on the number of tracking areas served by the base station. The components of a single VNF are instantiated on a pool of VMs in a 1:N mapping fashion.

Network services must be monitored in real time to automatically guarantee service continuity. A challenge in implementing failure detection and mitigation schemes is to ensure that alarm monitoring telemetry does not itself overwhelm the system. Hierarchical failure detection methodologies can be implemented to address this issue [12].

#### 8.5.2.2 Live Migration of Network Functions

Service continuity requires that the offered services are not negatively impacted by VNF scaling or reallocation. VNF portability enables VNFs to be migrated to guarantee service continuity. Live migration is the process of migrating VNFs from one host to another while guaranteeing zero or minimal impact to the connectivity service offered to mobile users. Being able to live migrate VNFs offers several significant advantages, including achieving load balancing across the available compute nodes by timely redistributing VNFs to sparsely loaded servers or reducing overall energy consumption by turning off a subset of the compute nodes once their VNFs have been moved elsewhere.

## References

[1] https://www.itu.int/en/ITU-R/study-groups/rsg5/rwp5d/imt-2020/Pages/default.aspx.

[2] Liolis, K., A. Guertz, R. Sperber, et al., "Use Cases and Scenarios of 5G integrated Satellite Terrestrial Networks for Enhanced Mobile Broadband: The SaT5G Approach," *International Journal of Satellite Communications and Networking,* Vol. 37, No. 2, 2019, pp. 91–112.

[3] https://www.ietf.org/about/2.

[4] https://www.itu.int/en/about/Pages/default.aspx.

[5] https://www.o-ran.org/.

[6] https://www.smallcellforum.org/.

[7] https://ofinno.com/article/multiple-transmission-reception-architecture-5g/.

[8] https://www.rfwireless-world.com/Tutorials/5G-NR-network-interfaces.html.

[9] https://www.prnewswire.com/news-releases/5g-services-in-north-america-300979188.html.

[10] https://5g-ppp.eu/.

[11] Taleb, T., A. Ksentini, and B. Sericola, "On Service Resilience in Cloud-Native 5G Mobile Systems," *IEEE Journal on Selected Areas in Communications*, Vol. 34, No. 3, March 2016, pp. 483–496.

[12] https://wwwetsi.org/deliver/etsi_gs/NFV-REL/001_099/001/01.01.01_60/gs_nfv-rel001v010101p.pdf.

# 9

# Building the vRAN Business: Technologies and Economical Concerns for a Virtualized Radio Access Network

## 9.1 What Are the Costs and Opportunities of 5G?

This chapter expands on the work in the previous chapter and centers more on the business considerations in the virtualization journey that is 5G. This includes consideration for the design and the various splits that are possible in the network. While there are technical options available, overarching concerns related to the business and economic models may prevail in many cases.

The business-side decisions will be centered primary around a cost model; both CapEx and OpEx. This includes the expenses to license spectrum, and deploy, operate, and maintain the network. These expenses need to be offset by the revenue generated from users.

Operators have relied nearly exclusively on the ARPU from the individual mobile consumer (even if bundled to enterprise accounts) to generate the revenues to pay for building and operating the network. The investments necessary to continually grow the capacity of the network will require revenue from additional sources, which directly implies expanding into markets that are underserved today.

In various jurisdictions, the licensing costs of the public spectrum alone can amount to many tens of billions of dollars investment or more for the CSP. With each generation in the mobile network (2G, 3G, etc.), there has been an allocation of new spectrum. While some previously licensed spectrum has been repurposed from 3G for use in 5G, there was a significant amount of newly allocated spectrum that required new licenses by the operators [1]. These licensing costs must be recovered by the operator during the expected operational life of the technology, usually less than 15 years. In addition to recovering the cost of entry into the mobile market by licensing spectrum, the costs associated with the procurement of the equipment and associated software licenses (which in many cases far exceeds the hardware costs) must be factored into the business model. These fixed costs are typically amortized as capital expenditures, although there is a trend starting to emerge where vendors are considering a pay-as-you-go model, and soon there may additionally be compute models that have CPU core license enablement as an option. In this latter case the CPU vendors may be considering a model where they will sell a base CPU with some limited number of CPU cores operational and with additional cores available on the die that can be enabled after manufacturing and procurement by the end user. In this model it may be possible to enable more compute resources on an already installed DU or CU simply by licensing more cores. Time will tell if this develops into a market option. One key consideration is presuming that the equipment capacity (e.g., the installed NICs and their configuration) along with memory or any other possibly limiting factor is capable of supporting the additional CPU core enablement without impairing other aspects of the total system, for example power consumption. This may be an interesting area of investigation for the curious system engineer.

The ongoing or recurring costs of power, maintenance fees (for hardware and software), and the operational team labor costs are typically associated with recuring operational expenditures. In response to these costs, one concern that is often heard from operators today is how can they monetize their 5G investment. With many markets at or in some cases beyond the saturation point of individual consumer subscribers, growth in the total subscriber base is unlikely to refill the coffers of the operators, and stiff competition in the market prevents nearly all operators from raising the monthly subscription fee (the ARPU). One business model falls into the zero sum game, where subscribers move from one carrier to another, and while this may benefit a specific operator over a competitor at any one given point in time, history has tended to show that this is a short-lived advantage, and over time there is some level of equilibrium in the market that remains relatively constant. This ignores the introduction of a truly new operator into a market. We previously mentioned a few examples where truly new operators have emerged: for example, Dish in the United States, Rakuten in Japan, and Reliance Jio in India. Yet here again, this mostly just moves the base of subscribers from one operator to another and really does not grow the overall total revenue in the community. Another potential source of new revenue is to expand the business, for example by providing managed wireless networks to enterprises. This has proven difficult to date for several reasons, many of which are not technical but rather are related either to the business model of the various enterprises, or are due to regulatory compliance concerns. In Chapter 12 more treatment will be given to the efforts going on in private wireless.

Setting aside the consideration to increase revenue by increasing the monthly subscription fee of the user community, the other knobs to turn in the financial model are to evaluate the TCO and reduce the various element costs found in either the OpEx or CapEx portions of the TCO.

Various TCO models are used both by operators and vendors in assisting in the technical decision-making process. Most of these models are very tightly controlled and business-sensitive due to the competitive nature of the business both for the operator and the vendor community. Nevertheless, a discussion on the basic elements can lead to significant insights.

As discussed in the first two chapters, one of the primary drivers of the NFV white paper was to change the way operators build and operate their networks, which was in response to the exponential growth in the data rate the network was being expected to support.

The deployment design choices, for example whether to consolidate CU/DU, are factors to be considered when sizing the capacity and number of servers needed. This in turn will impact the power consumption at different points in the network, and any increase in the number and locations of nodes to manage may impact the OpEx (staffing demands).

Likewise, the introduction of 5G into an existing (brown field) network may impact common elements deeper in the network, which is due to the anticipation of additional traffic through the network based on the availability of greater end-user capacity at the UE. This opens the aperture further to consider the elements in the TCO that are not specified by 3GPP. These include top of rack switches, core routers, and optical transport elements, all critical nodes in the network, none of which appear in any of the previous architectural figures. For example, in Figure 9.1 the interconnecting transport devices are conveniently omitted.

## 9.2 The 5G Business Outcome

The revolution that is virtualization is about the business outcome more so than the technology in many cases. This certainly is true for the transformation in the way operators are building and expanding their 5G and future 6G networks. The history of this evolution began with the addition of adjunct systems to legacy Class-5 switches, for example the AT&T 5ESS [2]. Note that here the word "switches" is not referring to L2 IP switches; rather, they are

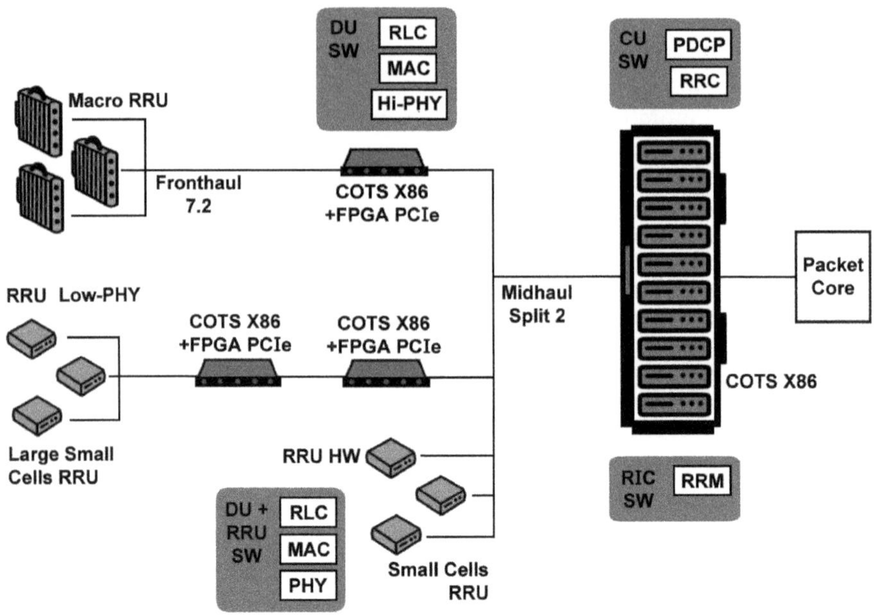


**Figure 9.1** 5G vRAN elements without the interconnecting transport devices.

public switched telephone network (PSTN) systems. With the earliest of the Gs, a network operator would augment their Class 5 switches with radio gear to interoperate with TDM circuits on the end office switch. The standard voice circuit of the day was capable of a maximum of 64 Kbps (in the United States and other jurisdictions that followed the ANSI standards, this was reduced to 56 Kbps due to bit robbing to enable a synchronization and signaling channel in the higher T1 framing). These systems would typically report metrics on a 15-minute interval. This data may include the number of call attempts, the number of call completions, the average call hold time (in some cases), and possibly peg counts (an incremental counting of some event) of some limited number of release codes. In addition, one might also see maximum CPU and memory utilization included in these reporting intervals. This data at the time was used as an indication of the health of the system and operational status of the node. For example, an increase of a certain set of release codes may indicate to the operator that calls were failing for some reason. Back-office systems in the OSS would often process these statistics and with some thresholds established raise alerts to operational staff if a threshold was exceeded (or a lower bound was breached). Clearly these statistical metrics of the day are outdated in today's network and shortly we will introduce and discuss the addition of the RIC in the 5G network.

In these early days the uplink capacity and the downlink capacity were identical. Soon this would transition into a model where the downlink capacity would far exceed the uplink capacity. We see this similarly in the wired network today. This brings a potential concern for the IoT and enterprise users where workloads such as streaming video will place higher demands on the uplink capacity. In Chapter 12, private 5G is discussed, including use cases that may place pressure on the asymmetry of today's designs.

The symmetrical nature of the network persists in the core of the network because of the basic end-to-end flow capabilities of what some are now calling the "donut" view of the network. Anyone familiar with running a speed test either on their wired network or on their mobile device will notice that the uplink speed is often a fraction, and sometimes a significantly smaller fraction, of the downlink speed. This of course is by design and reflects the evolution of the data flow of these devices; that is, the volume of video and the frame and bit rate from the network to the average UE is significantly greater than that of video flowing form the UE toward the network. As a result, if the network, specifically the RAN, were to have a fully symmetrical design today, a significant amount of capacity from UE to the network would be stranded or otherwise not used. Thus, the network has been designed to support the expected traffic demands of typical human users.

## 9.3 New Models to Address the TCO

Rakuten took the business TCO model to new levels when they began building their green-field 4G network and they continue to do so with their work in 5G. They staked a claim to be the first to implement a completely virtualized network. While this claim may have some conditions, it has proven a catalyst and serves as reference point for the operational transformation enabled by virtualization. Rakuten built a portion of their network with technology from the vRAN provider Altiostar and in August 2021 [3] acquired the company. In addition, Rakuten has recently undertaken an effort to license not only the Altiostar vRAN, but their entire solution to other operators, resulting in an innovative initiative to monetize their design through licensing to other operators. This development while still fresh is a new take on the model of the world's leading carriers building their own network gear (such as the Class 5 switches discussed above) and selling them into other markets. Indeed, the model of a handful of carriers setting the standard for design and deployment remains a common characteristic today with the second wave of operators following the lead of a few.

At the time of writing, Verizon is leading in the United States with their rollout of vRAN based on Intel servers with acceleration capabilities and they will soon introduce acceleration features native to the CPU. This model allows Verizon to run the BBU of the 5G RAN on standard high-volume servers (SHVS) [4].

The stated goal at Verizon is to deploy more than 20,000 cell sites by the end of 2025 on a virtualized platform, following the Open RAN model for disaggregating the RRU, DU, and CU accompanied with improvements in automation of the network configuration such as adding dynamic resource scaling based on demand. Verizon envisions a more responsive and robust 5G network that exceeds the capabilities of a traditionally built network. While the TCO is important, reducing the initial investment is not the leading factor for the innovative leaders in this space. Rather, the agility and improved operational efficiency of the network is the leading driver at this time, followed by the possibility of opening innovation. For example, in addition to deploying a virtualized RAN Verizon is also adding the capability to introduce MEC into the solution. The MEC servers in this case are likely to be deployed nearer the core (e.g., near the UPF systems) than the RAN units located remotely (e.g., at the base of the tower). There are several factors to be considered; one is the power and cooling requirements, along with space considerations at the tower site. These MEC systems in some cases are based on the AWS Wavelength product portfolio, and in the Verizon case, brings over 75% of the U.S. population to within 250 km of a 5G Edge node.

This model of disaggregation ideally differentiates vendors providing the software from to those providing the underlying hardware, which is typically based on SHVSs, also known as COTS servers (more on these servers shortly). While we have introduced the term SHVS here, the term COTS may appear from time to time in some literature. Note that they both mean the same thing.

## 9.4 The oRAN Model Introduces a RAN Intelligent Controller

The RIC is one of the possible innovation areas where CSPs may distinguish their 5G capabilities from the competition, or at least produce a competitive advantage by operating their network at a lower cost overall [5].

The development of the specifications in oRAN, and the associated disaggregation that is required (e.g., the decoupling of the system engineering) raises operational concerns.

Purpose-built systems of previous generations integrated the collection and management of operational data from various layers of the stack, from the hardware, Layer 1, 2, 3, and beyond, including the application layer, into what can be likened to a single pane of glass. This is often called the systems engineering of the solution. In the early NFV vision the EMS would close this gap. Unfortunately, that is not what has been evolving. The work in oRAN recognized this and the intent is to close this gap with the creation of the radio interface controller. A significant part of the intended functionality of the RIC includes SDN functionality originally envisioned by the 2012 NFV white paper (as discussed in Chapter 1). The SDN function in the RIC would enable the operator to optimize their network configuration in near real time, reacting to increases or decreases in the demand on the network. The RIC is separated into two functional elements (see Figures 9.2, 9.3, and 9.4) and would be deployed at separate locations within the network. One function is the real time (the RT-RIC) and the other is the non-real time (the NRT-RIC) function. Engineers are historically known for their lack of creative marketing names; this is clearly demonstrated in this case. The macro difference in the RT-RIC and NRT-RIC is segmented around processing and reacting in under one second for the RT-RIC and greater than one second the responsibility is the NRT-RICs. The RT-RIC could operate with a decision-making and action invocation in the 1-ms range for some functions, such as power management of the RAN

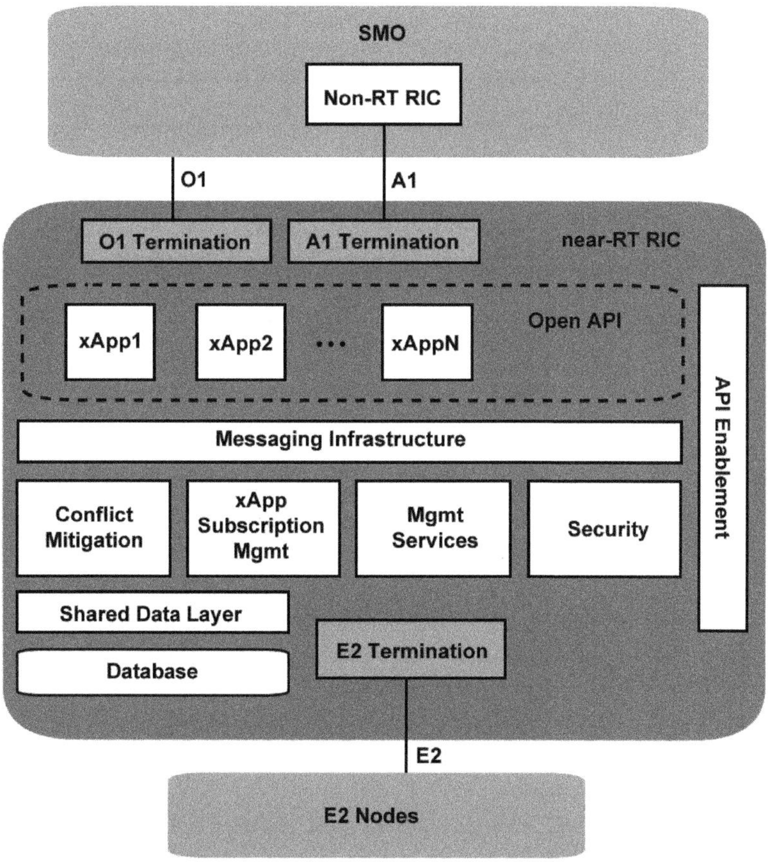


**Figure 9.2** The oRAN RIC.

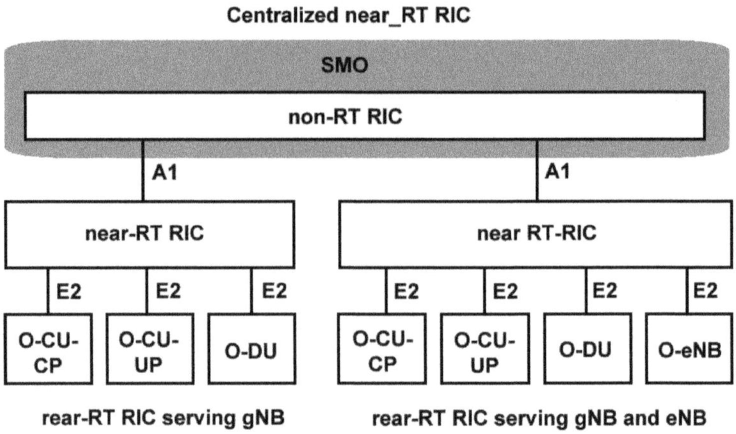


**Figure 9.3** Centralized near-RT RIC.

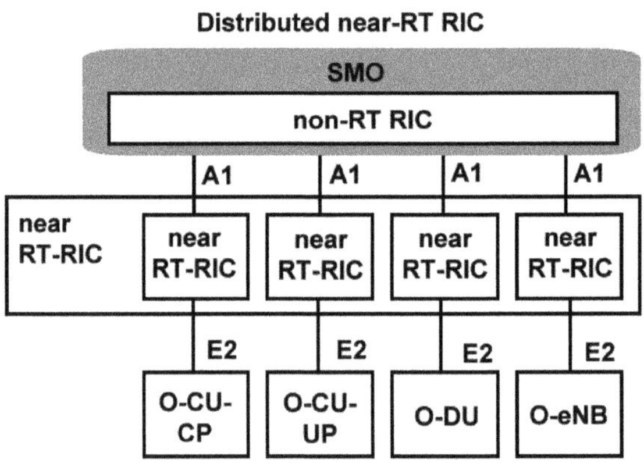


**Figure 9.4** Distributed near-RT RIC.

(amplifiers and CPU utilization), which will be discussed in more detail shortly. The primary stated function of the RIC is to provide SDN functions to the 5G RAN, more specifically to the open 5G RAN. The RIC in theory is going to be capable of providing AI or ML models to improve management, and the NRT-RIC may be the training location for the models. AI and ML are expected to enable more effective network operations by reducing human intervention and monitoring. Additional areas of management provided by the RIC include managing radio connections, QoS management, mobility management, and radio resource management. Central to this is the collection and analysis of analytics from the servers themselves. The modern CPUs upon which the oRAN are being implemented produce an astonishing amount of performance data that is often ignored at the application layer. One observable reason is the underlying host operating systems either supply their own metrics, or in some cases the metrics processing software is delayed until long after the release of the CPU. Another reason is many cloud workloads simply have little need for the granularity of the performance metrics that are collected, or the need for cloud-native systems' interoperability across various vendors CPU products makes supporting dissimilar functions from different CPU vendors difficult to justify, implement, and manage.

In the earlier Gs deployments, a vehicular drive analysis would be performed at least initially to measure and collect RAN performance data over the areas covered by the cell sites. This was costly in time and resources, and in addition the data sample was static (it was only valid at the time the sample was taken). The operator, or more likely a third party, would perform an actual vehicular drive in and around the area served by the cell site radios. This vehicle would collect a variety of performance data that once analyzed may result in a tuning of parameters for the particular cell site. One such tuning might be the actual cell site radio frequency or sector adjustments to improve power, reduce interference between adjacent cells, and other QoS functions. Not only was this exercise time consuming, but months or years may pass before any particular site could be reevaluated. Today, with the advent of the RIC for oRAN, and the ability of the UEs to also report operational data directly, the management of the QoS for the RAN can be adjusted dynamically. For example, if a new building is constructed in the footprint of the cell site, the RAN would be able to adjust to the new environment without any new drive testing and new analysis being performed. Similarly, the introduction of a new RAN into the same geography from a different operator would be easier to address, all automated through the analytics available from the RIC.

One of the more ambitious goals of the RT-RIC is the implementation of power management with a responsivity in the 1-ms range. In the oRAN model, where the DU is built on standard high-volume servers, modern x86 processors will accommodate P and C states [6], which allow the cores and noncore elements of the CPU to be placed in various suspended animation states. This allows for a very rapid reduction of power, and also a very rapid restoration of the operating state of the CPU. This capability enables the RT-RIC to command the CPU to respond quickly to the dynamics of the traffic load on the RAN. Previous models of traffic tracked busy hours, which allowed the RAN to enter lower power states but where the entry and exit of the reduced power states took considerable time. With the RT-RIC, a modern CPU with advanced power management and with enablement from the vendor ecosystem, the operator will, in theory, be able to track and predict microchanges in the RAN workload and transition into and out of various low-power states as the traffic pattern ebbs and flows throughout the day, minute by minute or second by second, possibly down to the microsecond level in future implementations. Operators also envision the potential for NRT-RIC analytics to create competitive opportunities to outpace their competition with improved QoS or QoE. These areas along with others around utilization of the available performance metrics are likely to remain active areas of future investigation and enhancements in the network for the foreseeable future [7].

## 9.5 Features of the One-Socket Server

The common data center computer today will typically have two to four sockets populated with CPUs. These servers, while well suited for cloud workloads, are not well suited for typical workloads and demands found at the DU and possibly of the CU of the 5G network. Extensive analysis based on the TCO modeling has shown, through private investigation, that high core count single-socket servers are preferred over multi-CPU servers. One area where the single-socket server has an advantage is in the lack of, or at least significant reduction in, concerns over NUMA alignment. The impact of NUMA was discussed in Chapter 5, but as a quick reminder, this is in essence the I/O (e.g., the NICs) must be aligned with the PCIe interfaces of each socket for maximum performance and managed latency and the workload of the process may have to be carefully designed to not break NUMA alignment.

As seen in industry publications, two leading network equipment providers, Nokia and Ericsson, currently have differing views of the current state of the virtualized RAN that bring different perspectives to the evolution of the RAN ("Nokia Ties Software to the Open RAN Maturity" [8], and "Ericsson Still Sees Flaws in Open RAN" [9]). However, one common point is that over time, the gap in performance between purpose-built RAN and an Open RAN implementation will be closed. On this point, the specification, software, and ecosystem will continue to evolve. Performance will continue to improve as CPU silicon incorporates features that allow for acceleration of the RAN workload and system integration of the disaggregated components becomes more streamlined (e.g., the RIC becomes a reality and becomes common to all the RAN vendors). Software vendors' adherence to standards, which do take some time to mature and become stable, will eventually lessen the interoperability challenges being faced today. Some of these challenges have been discussed previously—workload placement that avoids NUMA alignment issues is one example. Another example is where operators may mandate different middleware vendors or release sets and require the upper-layer software provider to adapt to their specification, such as the green field operator, Dish, which has a very ambitious goal of launching a fully cloud-native open RAN solution. Unfortunately, their plans have been delayed in part due to the challenges of integrating the various vendors' contributions to the overall solution.

## 9.6 Open Source Remains a Critical Element to the Virtualization Effort

Open source is one of the areas where the ecosystem sees value being taken out of traditional purpose-built systems. Lest we confuse this point, there is a cost to open source; open source does not equate to no cost. The cost is often borne in the system engineering to put together complete systems or the talent pool cost of maintaining a system built with open source. A 2020 study from *Infoworld* shows that the top five contributors based on the number of individuals who use their company email address are, in order: Microsoft, IBM (including Redhat), Google, Intel, and Amazon [10]. These companies are in business to make profits, and yet they make (through the labor renumeration of their employees) substantial contributions to the open-source community. One may reason that it may be easier to see how the first four in this list recover these costs through their paid products. The last one, Amazon, may have a more challenging time renumerating their investment; nevertheless, we see that they make a substantial contribution to the open-source community. Let's have a bit of a look at how this community developed to the state it is in today. Additional open-source discussion can be found in Chapters 10 and 13.

### 9.6.1 Open-Source Community in the RAN

The foundation of the open-source community lies with the kernel of the Linux operating system, which has been discussed as the underlying OS in the virtualization of the mobile network, and it has grown from the initial groundwork was laid down in the early 1970s [11]. Thirty years after the first release of Linux, the open-source community has nearly 15 million active developers, with GitHub itself reporting that there are over 200 million open-source repositories, and over 80 million developers, including contributions from over 4 million organizations [12]. There are many open-source projects and numerous testbeds available to the community, which will be discussed more in Chapter 13. These resources may serve as a reliable starting point to expand the working knowledge of the network elements, related protocols, and performance considerations.

## 9.7 Asymmetry in 5G and the Previous Gs

The current 5G standards and all the previous Gs, at least back to 2G, have been built around an asymmetric traffic model where the UE consumes (significantly in the modern era) more data than it uploads. This is driven by the way we as humans tend to use these devices—we stream video, which consumes a significant amount of bandwidth, and rarely upload more than a few seconds of video. There is some discussion in early phases that in 6G where the human to UE is not as significant a device in the overall traffic pattern; rather, sensors and in particular IP cameras and possibly vehicle sensor data and other high-data devices, may be the most prevalent users of the network. In this case it is highly likely that the UE will be a source of significantly more data being sent upstream than what we see in today's models. If this transition comes to pass, the power and traffic models of the RF along with other core functions will have to accommodate this changing model as well. Specifically, if IP cameras are considered, even with throttled or event-triggered traffic, steering the upload bandwidth of a few dozen cameras would saturate today's 4G base station. This may drive further use for enterprises to rely on edge or MEC computing resources at the enterprise edge rather than in the carrier network.

## 9.8 5G Market Drivers in Asia

The market drivers in greater Asia and in particular in China are at a scale like no other jurisdiction. There are two significant considerations that contribute to this scale: population and landmass [13]. China has clearly rolled out 5G at scale for the consumer market and continues to do so at a torrid pace. The total investment is nearly 60 billion USD in the RAN rollout alone and is forecast to approach 180 billion USD by 2025 with nearly 3.65 million 5G base stations. By contrast, mobile operators in the United States had reportedly deployed more than 100,000 by mid-2021, according to S&P Global Market Intelligence [14].

A market of this size should naturally catch the attention of the largest ecosystem partners yet the NEP landscape in China has proven difficult for some vendors; a report about Ericsson [15] makes this point clear. While the headline is more drastic than the actual impact, the pressure nevertheless is present. China has embarked on an indigenous project to transform from a consumer of high-tech silicon and solutions into a world leader, and a significant share of that technology will find its way into the CSP marketplace today. The leading NEPs in China are Huawei and ZTE as the two largest, with Huawei by far the dominant player. Private 5G in China is currently taking an approach where the CSPs maintain the RF and provide the services to the enterprise customers. By one estimate there are over 5,000 enterprises using what is termed "private 5G"0 in China today under this model. While the purists may choose to view this as an application of network slicing rather than a private network play, the use in enterprise should be viewed as more about hosting the workload and party to a SLA than about the license holder of the spectrum.

## 9.9 Business Considerations of Virtualization

Virtualization at the end of the day is about changing the business model of how the networks are built and operated, and in doing so may introduce the possibility that the CSPs can innovate their business offerings. This should directly lead the carriers to monetize the enterprise space, and yet this market has been slow to mature. A recent study out of the United Kingdom [16] reports that industrial 5G and industrial IoT are one and the same, and more significantly that while the market is massive in general, the specifics are small when applied to individual applications. This observation indicates that what is often called the private 5G or private wireless market is fragmented. The ability to scale any one application or set of applications to meet the cost models used by operators when allocating their capital remains a challenge. Where the operators are engaged, they seem to focus on a very limited set of verticals, struggle with the various use case business models, rely on partners, or choose not to pursue some business use cases.

The industry has yet to converge on a standard product model that can meet the needs of the large number of different enterprises with the same scale that the operators have managed to serve in the consumer space. In Chapter 12 we will take up the hyperscaler discussion and consider where they may have more agility to address these markets and are moving rapidly to fill the gap the CSP's are not serving.

## 9.10 Pro and Cons of White Boxes, Which Are Truly SHVSs, in the vRAN

Nearly all our network models have shown a direct connection over the interfaces from one node to another. While this is valuable for understanding the discussion provided, it ignores the actual physical network connectivity in many cases. These nodes (e.g., BBU, DU, CU, and UPF) are all likely not single servers directly connected to each other; rather the RRU may actually be multiple RRUs connected to a DU server or servers, which implies that there are TOR switches at the base station locations. In addition, there is often a cell site router (CSR) at the base station location that consolidates the traffic for the 5G, 4G, and any other provided Gs at that side before the CSR interfaces to the fiber connection (presuming there is fiber at the tower). There are cases where there is not a long-haul fiber connection to the base station. In these cases, there may be microwave connection to a nearby base station or other node where we would find a fiber drop. Nevertheless, the fronthaul fiber has a long-haul capability where the transceiver in a CSR is capable of many kilometers of light transmission. Once we reach deeper into the network, we find that the CU or other elements are also colocated with other compute nodes, possibly in the same rack. Here we again find TORs connecting the various compute nodes to the network. This model continues as we progress further into the network. These switches historically have been provided by large system vendors on purpose-built branded platforms. One of the adjacent efforts associated with the virtualization of the network is to introduce white-box switches to replace the vendor-specific purpose-built TORs. By now we would expect that the discussion of a white box is easily overloaded in use and discussion, and one would be correct in thinking that this is the trend. The industry has overloaded the use of terms like bare-metal switch, bright-box switch, and white-box switch and permits that the context of the discussion or use, when necessary, allows the reader to discern the difference. We will provide some segmentation of the various instances here, then revert to simply calling them all white-box switches. A bare-metal switch consists of the hardware elements up to and including any firmware required to bring the device to a load point ready for an installed operating system to boot. These bare-metal switches are often available from original equipment design (ODM) vendors.

## 9.11 Bright Boxes: Standard High-Volume Servers with One or Two Customized Features

Several efforts have been made to describe modifications or enhancements to SHVS found in many hyperscaler data centers today. These servers are often available from a number of vendors and are specified by the hyperscaler. In many cases these servers come from ODM vendors and are unbranded. The hyperscalers have several advantages due to their unique market size. One is that they can specify very specific configurations and contract directly with the actual manufacturer for the servers that are deployed in the millions in their data centers. In addition, the hyperscalers have designed their systems with a resiliency model where if any one server suffers a physical failure—for example the memory becomes defective—that server is simply removed from service, and others in the vast data center are fully capable of handling the workload. At times—for example when workloads require advanced processing as might be found in AI applications, either in the inference or the training of the models—the hyperscaler may require additional hardware accelerators. These hardware acceleration cards are additions to the SHVS and may be added by the hyperscaler after the servers reach the hyperscaler. These accelerators can be GPU cards for training and inferencing, FPGA cards for a number of interesting workloads including training, and other functions. These accelerator cards convert a SHVS into a bright-box server. The use of this term may be diminishing slightly but still appears from time to time in some discussions. Simply put, it can be considered as an adaption of a SHVS with some acceleration features, likely by the addition of a PCIe card or two, that is not native to the core CPU, networking, or storage functions of a standard server.

## References


[1] Manner, J. A., *Spectrum Wars: The Rise of 5G and Beyond,* Norwood, MA: Artech House, 2022.

[2] https://ieeexplore.ieee.org/document/6772422.

[3] https://www.altiostar.com/rakuten-group-to-acquire-mobile-industry-innovator-altiostar/.

[4] "Verizon Deploys More than 8,000 VRAN cell Sites, Rapidly Marches Towards Goal of 20,000," https://www.verizon.com/about/news/verizon-deploys-more-8000-vran-cell-sites.

[5] "5G RIC – RAN Intelligent Controller" https://www.techplayon.com/5g-ric-ran-intelligent-controller.

[6] https://www.intel.com/content/www/us/en/develop/documentation/energy-analysis-user-guide/top/energy-analysis-metrics-reference/p-state.html.

[7] "5G RIC RAN Intelligent Controller" https://www.techplayon.com/5g-ric-ran-intelligent-controller.

[8] https://www.sdxcentral.com/articles/analysis/nokia-ties-software-to-open-ran-maturity/2022/09/?campaign_uuid=54da3230-72e7-4b15-a5d5-712f45a338bd&member_uuid=97dcb160-b632-4db3-afae-e224cc3efbb7&newsletter_event=pageview&utm_campaign=2022-10-04&utm_content=ran-topic&utm_medium=email&utm_source=newsletters.

[9] https://www.mobileworldlive.com/featured-content/top-three/ericsson-still-sees-flaws-in-open-ran/?ID=a6g1r000000zRzaAAE&JobID=1257328&utm_source=sfmc&utm_medium=email&utm_campaign=MWL_20221006&utm_content=https%3a%2f%2fwww.mobileworldlive.com%2ffeatured-content%2ftop-three%2fericsson-still-sees-flaws-in-open-ran%2f.

[10] https://www.infoworld.com/article/3253948/who-really-contributes-to-open-source.html.

[11] "The History of Linux," https://linuxhint.com/history-of-linux/.

[12] "Let's Build from Here," https://github.com/about.

[13] Tomas, J. P., "Chinese Carriers Deploy Over 1.97 Million 5G Base Stations," RCR Wireless, August 19, 2022, https://www.rcrwireless.com/20220819/5g/chinese-carriers-deploy-over-1-million-5g-base-stations-report.

[14] S&P Global, "5G Tracker: 79 Markets Worldwide Have Commercial Services," February 24, 2022, https://www.spglobal.com/marketintelligence/en/news-insights/research/5g-tracker-79-markets-worldwide-have-commercial-services.

[15] Morris, I., "Ericsson to Cut Hundreds of Jobs in China After 5G Setbacks," *Light Reading,* September 29, 2021, https://www.lightreading.com.

[16] Blackman, J., "Operators Upended by Hard Graft of Private 5G–A Classic IoT Tale, Stuck on Repeat," *Enterprise IoT Insights,* October 11, 2022, https://enterpriseiotinsights.com.

# 10

# Designing Virtualized 5G Networks

## 10.1 Successfully Designing Virtualized 5G Networks

As discussed in this book, virtual systems and networks are proving to be extraordinarily successful due to their appealing features. However, transferring the virtualization concept into real-world systems and designing practical network solutions are not trivial endeavors. The separation of hardware and software requires a new and different engineering thinking compared to the way previous generation of mobile networks were designed. Moreover, the disaggregation of functions introduces an additional new design dimension for the engineering of the network.

Hence, the question arises as to when the engineering of a virtualized system or network reaches a useful and successful implementation or instantiation of the target? This important question can be reformulated as follows: How does a concise and well-founded engineering of the system contribute to its economic and technical success?

The answer to this question is highly complex. However, attempts must be made to address this point while accounting for different engineering concepts. This chapter aims to present a broader view on how successful engineering of a virtualized 5G system can be achieved and which engineering techniques should be used. It does not discuss general engineering concepts such as agile [1] or DevOps [2]. Instead, the chapter aims to identify a mapping between a successful system, the required capabilities, and the required performance with methods and more detailed concepts and methods from software, software engineering, data structures, and computing concepts and architectures. Particular attention is given to open-source software (OSSW) and its reuse along with a discussion of performance criteria for 5G systems.

### 10.1.1 What Is Success for a Virtual System Design?

Success is typically the accomplishment of an aim or purpose. This notion will be interpreted in the coming section in the context of virtualization, cloudification, and disaggregation.

### 10.1.2 Overall Aim

Foremost, the system theory for virtualized systems outlines that these structures should be nearly indistinguishable from a possible physical implementation of the same purpose. This view emanates from the "fidelity" or "equivalence" requirement for virtual systems as formulated, for example by Popek and Goldberg in 1974, and already discussed in Chapter 3.

#### 10.1.2.1 Network Performance

The importance of this equivalence requirement must not be underestimated. While the fidelity requirement was initially interpreted as achieving the same computational results, virtual 5G networks require a more modern and forward interpretation of this aspect. The most obvious obligation is that virtualized 5G must achieve the same system behaviors and, particularly, the timing and performance as their nonvirtualized and nondisaggregated counterparts. The systems not only need to achieve the same goal but also must not differ in their temporal behaviors as seen from an external point of view. We will outline the performance requirements later in this chapter and focus next on successful virtualization.

### 10.1.3 Efficient Virtualization

Investigating the origin of the fidelity requirement for virtualization concepts may help illustrate an important success criterion for virtual systems design. The early virtualization systems were designed to share resources among multiple users or processes (i.e., the access to expensive CPU, disk storage, or main memory). A noncoordinated access to the memory, for example, permitted processes to overwrite each other's data, leading to corrupted data being used in computation and generation of an incorrect result. Hence, a simple but successful virtualization technique for memory must at least be able the efficiently coordinate (a) the access timing (i.e., coordinating the temporal order), and (b) the choice of physical location (i.e., spatial behavior) of the system.

The relevance of the temporal alignment between requests for resources of virtualized system was outlined in Chapter 5. Section 5.2 introduced a metric to account for the change of requests' timing. On the other hand, the characterization of the spatial behavior is eventually more complex. The knapsack problem [3], which is a widely used and well-understood problem in theoretical computer science, may help clarify what kind of approach is needed in this case. The knapsack problem, colloquially speaking, is the task to fill a rucksack with as many items as possible without exceeding the weight limit of the rucksack. Note that the items might have various weights. In virtualization, the knapsack problem can be applied as follows: memory requests are the items, and the total amount of available memory is the weight limit. The knapsack problem can also be seen as a search procedure. Heuristic algorithms exist to solve this problem efficiently (e.g., to find in a short time an allocation that leads to a high utilization of the memory or any other type of resource).

### 10.1.4 Separation and Portability

The example of virtual memory clearly outlines the next capability that is needed in virtual systems, which is the separation of physical relationships from logical relationships. Formally characterizing the degree (or quality) of separation is probably the most challenging task in virtual system design. P2P networking research (e.g., [4]) has demonstrated that overlay networks can be more resilient than the underlying physical networks. However, the evaluation of the quality of resilience is currently only possible within the specific context of this objective (i.e., resilience of data transmission in this case) and unfortunately cannot be generalized.

An even more pragmatic definition of successful separation of physical relationships from logical relationships must be sought in another feature that is the functionality "portability." Virtual systems must allow functions to be moved to arbitrary locations and eventually be executed in parallel. Moving functions between various nodes, however, requires a mature level of abstraction and context in order to enable reuseability of the function at multiple locations.

### 10.1.5 Open-Source Software

Adding to this point, reuseability of software is another important feature in the design and engineering of software. As outlined in Chapter 3, the separation of software and hardware was developed hand in hand in the 1970s and is related to the development of virtualization techniques. The separation enabled new business models to emerge in computer system engineering. The decoupling and unbundling of software from hardware permitted the creation of companies that focused solely on developing software, thus enabling them to specialize in software functionalities. Not only did software development effort become somewhat decoupled from the speed of the hardware development, even more important, the resulting advances in software development led to an acceleration of the hardware development itself. A typical example is the development of improved operating systems that required hardware support for hypervisors, which in turn led to the implementation of this feature in modern CPUs.

However, the strong coupling that in some applications exists between software and hardware (i.e., software only being executable on specific CPUs) should not be underestimated. This characteristic, together with the general availability of more computing power, may enable another approach: the use of higher abstraction layers or translation mechanisms, for example, by containers or just-in-time compilers. Ideally, the mapping and transfer are automated for the latter (e.g., the orchestration of containers using Kubernetes has become very popular). However, the ultimate flexibility is achieved when human-readable software (source code and associated development and deployment aids) are able to be reliably put into production by others than the original developers; for example, the derived function resulting from the code can be instantiated in a production network.

The ability to read and understand code pushes the design of virtualized networks to the next level of abstraction: the use of open-source software. OSSW is computer software that developers have the rights to freely use (or nearly freely), study, change, and distribute to anyone and for any purpose. Open-source software is typically released under a license that the copyright holder grants developer arbitrary access.

Reuse of resources is a form of virtualization, and the following sections discuss the advantages of open-source software and review the open-source software packages that are available in 5G.

## 10.2 Open-Source Software for 5G

A number of open-source software initiatives for 5G and B5G systems have been announced recently. These initiatives comprise well-proven international standardization efforts within the telecommunications industry such as the ETSI-hosted Open-Source MANO (OSM), [5] project, that focuses on the network operators needs for software for the management and network orchestration (MANO) of 5G/B5G systems. However, additional open-source initiatives have recently gained traction such as cloud-native computing [6] and Kubernetes [7]. We will outline these solutions later in Chapter 13, the advantages of using open-source software in ICT system design will be covered in more detail next.

### 10.2.1 Why Open-Source Software?

OSSW can be seen as an extension of the software business that has its foundation in the 1970s when hardware and software businesses were first separated. Today, an appealing feature of the OSSW concept is to separate the development of software from the activities in the software lifecycle. OSSW enables software engineers to detach the basic design of a software module from adapting the software in a specific use case and to maintain it for this use case. Studies on the use of OSSW have clearly shown that significant software development cost reductions are achievable in general. These savings often originate from the advantage of reusing OSSW modules across a number of use cases, but also other reasons such as tapping into a larger domain-specific developer community. Next, we outline the various advantages of using OSSW in software development.

### 10.2.2 Flexibility and Agility

OSSW yields increased flexibility and agility in industrial software design and development. Flexibility is the ability of the design process to respond to uncertainty and still succeed. It comprises the openness to changes, assumptions, goals, or the course of the process. OSSW achieves flexibility by starting with a rather mature solution, which is subsequently open to changes that can be adapted and customized. Furthermore, OSSW may improve agility in the design process. Agility in this context means the discovery of requirements and software quality improvement through collaborative efforts in the design process. OSSW enables technology agility by offering multiple ways to solve problems. For example, OSSW may prevent a project design process from slowing down due to a particular capability that is not available. Instead of waiting for an outside-of-the-company developer to add the missing capability, the in-house designer can contribute the lacking capability. The efforts of an in-house development might be non-negligible but may become predictable and, in turn, might be communicated or charged to the customer. Another practical variation is the open-source community can and does implement features and enhancements that do not need a specific business justification (e.g., there is money to be made if a specific feature is implemented); rather, the technical community alone may decide on the technical value of a feature and this is sufficient for the feature to find its way into the code.

### 10.2.3 Speed of Development and Deployment

OSSW accelerates the software design by reusing software and enabling companies to improve their competitiveness. OSSW permits the design process to leverage community versions, understand whether the software addresses the technical and business objectives of the problem, and begin to deliver value right away. Moreover, professional support and services are increasingly becoming available for open-source products (e.g., by companies like IBM with their Red Hat product). This permits the software designer to focus on the values and functional features that the actual designer is an expert in.

Moreover, the combination of development speed, flexibility, and agility that are offered by OSSW permits innovators' solutions to mature rapidly to large-scale, fully supported, and enterprise-grade implementations. The use of OSSW permits a software developer to focus on and improve nonfunctional system features like scalability, reliability, and performance. While the implementation of a specific function is typically provided in the OSSW modules, the developer using the modules can in parallel improve the software systems' nonfunctional features without fearing that the provided function is corrupted [8, 9].

### 10.2.4 Low Licensing Efforts

OSSW software licenses are typically general (they are not individually negotiated or issued for a single specific use case). They simplify the relationship between software authors and licensee. OSSW efforts typically seek to contribute software to the public domain, like a Creative Commons license that is a legal tool for waiving as many rights as legally possible [10]. However, other public license models, like the GPL [11], guarantee end users the four freedoms to run, study, share, and modify the developed software, imposing certain commercialization requirement. While the latter options might add a limited amount of complexity in OSSW license management, OSSW licensing avoids the hurdle of propriety license specifications and negotiations that often block design activities and slow down development execution.

However, open-source code is not guaranteed to be free from licensing issues; for example, the code may contain algorithms or techniques that may impinge on the intellectual property rights of others who may require renumeration for commercial usage (voice codes are one such example).

There clearly are advantages to using open-source, and one must also acknowledge that there may be some disadvantages, some which we detail in Table 10.1.

### 10.2.5 Cost-Effectiveness

OSSW is generally much more cost-effective than a proprietary solution. OSSW-based software solutions use software components that are typically already at a high maturity level. The remaining effort of fine-tuning these components is relatively small compared to the overall effort of developing the OSSW components themselves. Moreover, the OSSW gives developers the ability to start

**Table 10.1**
Pros and Cons of Using Open Source

| Pros | Cons |
|---|---|
| • Bypasses centralized consensus, thus reducing time to implementation<br>• Fast bug fixes and improvements in most cases (some may argue otherwise)<br>• Community development of core functionality, lots of developer resources<br>• Low barrier to participation by smaller players; just go get the code and start working<br>• Strong participation by IT players, massive ecosystem | • Reduced implementation diversity increases vulnerability<br>• Lacks recognized/persistent governance authority—Who determines priorities for releases and cadence of releases?<br>• Unconstrained code contributions leading to bloat/bugs<br>• Feature persistence from release to release—What constitutes "normative" in open-source code?<br>• Risk of lock-in through forking<br>• Uncertain licensing implications<br>• Difficult to ensure security by design<br>• Culture: hard to manage individual developer contributions<br>• Vulnerable to fragmentation into multiple communities<br>• Stretches developer resources—lifecycle management can be an issue<br>• Poor/nonexistent documentation on what the code does or does not do |

small in their efforts and to eventually scale (see below). Given that development budgets are typically limited, it just makes financial sense to explore open-source solutions.

### 10.2.6 Ability to Start Small

When using OSSW, a software design can start small and be focused (i.e., it can address only the functions that are missing or target software specific modules that the developer is most experienced in). The newly developed functions can be turned into a community version, for example, for testing, and then migrate to a commercially supported solution, such as functions or services with very high reliability, as dictated by the business-specific requirements. If the project does not require support, then one can continue to contribute to the community version indefinitely with eventual mutual benefits. In this case, a designer has the option to try the various alternatives, pick the one that works best, and then scale up its implementation within a commercial solution.

### 10.2.7 Software Security

Commercial OSSW usually has a strong information security record. It is unlikely for OSSW-based solutions to claim security superiority that does not effectively exist because developers can continuously verify and validate the security claims made on the developed software. The software is not left to molder, unlike a proprietary development environment where only a few developers know how the code works and are aware of its flaws while being unable to overcome them due to resource or skill issues. An open developer community can identify and fix problems rapidly when they became apparent. Hence, in theory the overall security quality of the software increases.

In practice there is a risk that the theory does not hold absolute truth; one concern is developers may not have the skills or time needed to properly evaluate the security of the code. Often, it is only after a breach in the wild that the vulnerability is uncovered—the damage has been done by this time. An additional problem may arise when many different systems in a product network use the same open-source module, which may lead to a large-scale failure, a cascade of failures. Ensuring diversity in code implementations is one way to overcome this problem, but then the advantages of open source are diminished.

### 10.2.8 Shared Maintenance Costs

A fundamental advantage of OSSW stems from wide community involvement. Rather than writing the code for an application and having the application developer sustain the burden of maintaining it, the developer can share the cost of maintaining and sustaining the developed applications with multiple parties. This is usually a mutual process with complementary benefits. For example, the application developer can hand over difficult maintenance problems to the community. In turn, the community becomes aware of the new maintenance requirements and can find suitable approaches to overcome the hurdle. While the required software updates typically take time and incur significant costs, parallel development activities are possible in OSSW and can reduce completion time. One practical concern that can occur is either a forked branch is created and never merged back into the mainstream, or the proposed update is rejected for some reason by the maintainers.

### 10.2.9 Enabling Future Development and Attract Better Talent

Companies seeking software developers have an edge when it comes to attracting talented personnel. Most professional developers and software engineers are well-aware of open-source software projects and can become a company asset immediately upon hiring. They have developed their own skills to contribute to and take advantage of OSSW-projects (i.e., how to apply the software and how to deal with the common risks of using OSSW-based code). Many enjoy creating their own projects and having the ability to interact with other developers outside their enterprise environment. Giving developers flexibility and freedom is an important aspect when attracting the best talent as well as motivating current employees.

The increased motivation and creativity of the OSS developer often yields technical solutions that lead to new product features. Web, Cloud, and IP-based mobile applications are increasingly built on open-source infrastructure. Some data and analytic solutions are only available in open-source. Hence, future generations of mobile network infrastructures are highly likely to be based on open-source software solutions, at least in part.

## 10.3 5G Open-Source Efforts

This section provides an overview of open-source 3GPP software implementations and some deployment frameworks that can be used to build proof-of-concept (PoC) designs. In comparison to costly proprietary and closed-source 3GPP-compliant software, open-source solutions (when combined with readily available laboratory RF hardware) allow experimenters to build low-cost small-scale PoC designs that allow the use of commercial cellular terminals (user equipment) as opposed to emulators or non-standards compliant equipment. In addition, the open-source approach facilitates the integration with commercial (i.e., vendor) network elements. The latter proves useful in collaborative projects with industry partners, in addition to providing testing solutions for industry research and development labs.

### 10.3.1 Open-Source 5G Core Network Elements

The core network in a 3GPP 5G radio system constitutes the collection of protocols needed to interconnect radio-access networks and provide interoperable services such as telephony and internet access for mobile users. The main entities comprising the network include both control plane and user plane functions. We will defer until Chapter 13 a more detailed discussion of some of the popular open-source software packages that may be used to run 5G core networks that are significant and were previously in the discussion found in Chapter 9.

## 10.4 Design and Performance Criteria for Virtualized 5G Systems

Successful virtualized 5G systems must fulfill the same requirements as nonvirtualized ones. In addition, however, they must meet the operational and performance criteria of any typical successful virtualized system design. Hence, we postulate that successful virtual 5G systems must comprise both (a) the best practice and successful implementation of virtualization and software concepts along with (b) meeting the performance criteria of general (e.g., nonvirtualized) 5G systems.

The next section provides both labeling and mapping of software design concepts as they apply to the functions required in 5G systems. In addition, the section also describes the general performance criteria and performance requirements of 5G systems. The provided description of the required functions is based on the service requirements for 5G Release 16 [12], as available at the time of writing. The used labelling adopts the ontology of the Association for Computing Machinery (ACM) advised for subject classification for computer science.

### 10.4.1 Computer Systems and Software Engineering Concepts for Virtualized 5G Systems

5G systems are rather complex systems and are required to fulfill many functional capabilities. More specifically, they are essentially driven by the needs for (a) supporting very different types of user equipment (user equipment, for example, for the IoT), (b) new operational and usage scenarios, as well as (c) new novel network services and (d) the development of new networking techniques and concepts, such as SDN, and finally, 5G systems are required to (e) have very high performance, efficiency, and resilience.

The main drivers behind the 5G system design [13]:

1. Support for multiple access technologies;
2. Ability to scale in size and be highly customizable networks (e.g., the slice concept);
3. Advanced key performance indicators (KPIs), such as service availability, latency, reliability, and user experienced data rates;
4. Support for very-high-area traffic capacity;
5. Flexibility and programmability, such as network slicing, diverse mobility management, the concept of hardware and software separation, and disaggregation as in network function virtualization;
6. Efficient resource utilization, both in the user plane and control plane, but also at the infrastructure level;
7. Seamless mobility in densely populated and heterogeneous environments;
8. Support for real-time and non-real time multimedia services and applications with advanced QoE.

## 10.5 Computer Systems and Software Engineering Concepts for 5G Functions

Typically, these network characteristics translate to various methods that are related to the eight categories on networks concepts as suggested by ACM [14]: (1) network algorithms (e.g., data path or control path algorithms), (2) network architectures (e.g., layering or programming interfaces), (3) network components (e.g., end nodes, middlebox, or physical links), (4) network performance evaluation (e.g., modeling, simulation, or analysis), (5) network properties (e.g., security, dynamics, or reliability), (6) network protocols (e.g., protocol design or protocol correctness), (7) network services (e.g., naming and addressing, programmable networks, or network monitoring), and (8) network types (e.g., data center networks, WANs, or mobile networks).

The combination of network characteristics and network concepts spans the very wide range of the 5G networks design space. Moreover, systematic work in this design space has led to the definition of 5G systems as we know them. This design space may offer a framework value for the overall system engineering effort on the one hand, and high academic and scientific value for scoping the related research missions on the other.

Unfortunately, network characteristics and concepts alone do not provide detailed suggestions as to how computer systems concepts must be applied in the design of virtualized 5G systems. This deficiency is probably also related to coconsideration of hardware on one side, and functionality of software on the other. It is therefore necessary to provide a complementary design approach by considering the functionality of 5G and suggesting nonnetworking (i.e., software, software engineering, data, and computing) capabilities that are expected to be of the essence in the design when one uses virtualization and disaggregation in these systems.

These capabilities and requirements are also suggested in 3GPP/ETSI [15]. Due to the continuous changes of some of these capabilities (e.g., eV2X), high domain-specific (e.g., for cyber-physical control applications in vertical domains), or special markets (e.g., extreme long-range coverage in low-density areas) this chapter focuses only on those that are well established at the time of this writing.

Table 10.2 offers an overview of the specific functional capabilities for 5G systems, their requirements, and selected suggested computer systems and software engineering concepts that can address them. To best understand this table, the reader is encouraged to become familiar with the ACM classification [16].

While the content of Table 10.2 is expected to change over time, it becomes self-evident that additional performance and success criteria for the engineering of virtualized 5G systems are needed. The focus on software and abstraction in virtualized and disaggregated systems requires the increased use

**Table 10.2**
Concepts from Software, Data, and Computing to Address Design Challenges of Virtualized 5G Systems

| Basic Capabilities | Requirement | ACM Software, Software Engineering, Data, and Computing Concepts |
|---|---|---|
| Network slicing | • Functions to create, modify, and delete a slice<br>• Connectivity to home and roaming users within the same slice | Concurrent programming, visual programming, requirements/specifications, data structures, object-oriented programming |
| Diverse mobility management | • Optimize network's mobility management based on the mobility patterns | Data structures, concurrent programming |
| Multiple access technologies | • Enable the UE to select, manage, and efficiently provision services and to select a certain access network<br>• Dynamically offload traffic between access technologies<br>• Simultaneous data transmission via different access technologies<br>• Support UEs with multiple radio capabilities | Data structures, concurrent programming, process management, communications management |
| Resource efficiency | • Minimize control and user plane resource usage for data transfer<br>• Optimize the resource use of the control plane and/or user plane for transfer of small data units, including high data rate (e.g., 10 Mbps) and very low end-to-end latency (e.g., 1–10 ms)<br>• Support high-density connections | Performance, reliability, data structures, concurrent programming, optimization |
| Efficient user plane | • Set up and management of efficient user plane paths for UE and including support for:<br>  • Mobility<br>  • Service and application hosting environments located within or outside the operator's network<br>  • Maintaining user experience (e.g., QoS, QoE) | Concurrent programming, requirements/specifications, data structures, reliability, security, and protection |
| Efficient content delivery | • Efficient delivery of content from caches under the control of the operator (e.g., locate caches close to UE) | Concurrent programming, data structures, optimization |
| Priority, QoS, and policy control | • Flexible mechanisms to establish and enforce priority policies among the different services and users<br>• Provisioning of required QoS<br>• Support of E2E QoS for services, including hosting environments | Concurrent programming, requirements/specifications, data structures, reliability, security, and protection |
| Dynamic policy control | • Optimized signaling in real time for prioritized users and traffic | Concurrent programming, requirements/specifications, data structures, reliability |
| Network capability exposure | • Suitable APIs to allow a trusted third party to create, modify, monitor, and delete network slices<br>• Allow a third party to associate a UE to a network slice | Concurrent programming, requirements/specifications, data structures |

**Table 10.2** (continued)

| **Basic Capabilities** | **Requirement** | **ACM Software, Software Engineering, Data, and Computing Concepts** |
|---|---|---|
| Context-aware network | • Provide context information to support efficient network resource utilization and optimization, including:<br>  • Network conditions<br>  • UEs<br>  • Application's characteristics | Data structures, process management, concurrent programming, software architectures |
| Self backhaul | • Support NR- and E-UTRA-based wireless self-backhaul<br>• Flexible partitioning of radio resources between access and backhaul<br>• Support autonomous configuration | Data structures, process management, concurrent programming, organization, and design |
| Flexible broadcast/ multicast service | • Downlink only broadcast/multicast for a specific area<br>• Enable the operator to reserve radio resources | Data structures, process management, concurrent programming, software architectures |
| Subscription aspects | • Allow for identifying a UE as an IoT device<br>• Mechanisms to change associate between subscription and IoT device | Data structures, process management, software architectures |
| Energy efficiency | • Energy-saving modes that:<br>  • Can be activated/deactivated either manually or automatically<br>  • Can be restricted to a group of users | Data structures, process management, software architectures |
| QoS monitoring | • Supporting real time E2E QoS monitoring<br>• Different levels of granularity for QoS monitoring | Data structures, process management, software architectures |

of programming, architecture, data, and computing concepts in distributed systems.

## 10.6 Performance Criteria for 5G Systems

The performance criteria for 5G aim at characterizing the (a) performance of the system and service functions of 5G networks (see above), and at (b) end-user and perceived performance criteria.

They can be understood best if one considers the evolution from 4G to 5G systems. Figure 10.1 is taken from an ITU suggestion for a 5G system [17] and compares the capabilities of the IMT-Advanced, 4th generation, with the expectations for IMT-2020 (i.e., 5th generation mobile networks). The figure uses logarithmic axes that permits an easier identification of the foci on the general performance categories by 5G systems. These systems are expected to exceed the previous generation network, notably in capabilities for higher area

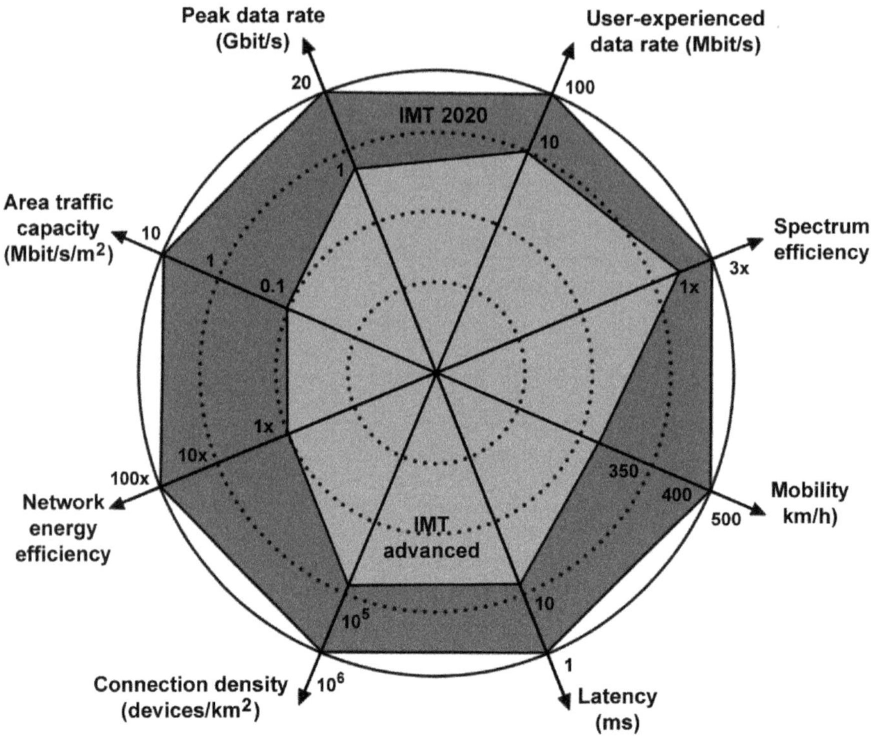


**Figure 10.1** Performance expectations for 5G mobile networks.

traffic capacity and for network energy efficiency. However, the improvements in peak data rate, user experienced data rate, and latency are impressive. But improvement in spectrum efficiency, and the support of higher mobility speeds by pushing the application of 5G system toward extreme mobility scenarios, is still needed.

The requirements for experienced data rates and coverage area traffic capacity are outlined in the Table 10.3, which was taken from Section 7, "Performance Requirements" in [12]. In addition to the data rate requirements, the ETSI document outlines several scenarios that require the support of very low latency and very high communications service availability. The requirements are often denoted as URLLC. However, the overall service latency depends on the delay on the radio interface, transmission within the 5G system, transmission to a server that may be outside the 5G system, and data processing. Moreover, some of these factors depend directly on the 5G system itself, whereas for others the impact can be reduced by suitable interconnections between the 5G system and services or servers outside of the 5G system (e.g., to allow local hosting of the services).

**Table 10.3**
Requirement Data Rates and Area Traffic Capacity by 5G Systems

| Scenario | Experienced data rate (DL) | Experienced data rate (UL) | Area traffic capacity (DL) | Area traffic capacity (UL) | UE speed |
|---|---|---|---|---|---|
| Urban macro | 50 Mbps | 25 Mbps | 100 Gbps/km$^2$ | 50 Gbps/km$^2$ | Pedestrians and users in vehicles (up to 120 km/h) |
| Rural macro | 50 Mbps | 25 Mbps | 1 Gbps/km$^2$ | 500 Mbps/km$^2$ | Pedestrians and users in vehicles (up to 120 km/h) |
| Indoor hotspot | 1 Gbps | 500 Mbps | 15 Tbps/km$^2$ | 2 Tbps/km$^2$ | Pedestrians |
| Broadband access in a crowd | 25 Mbps | 50 Mbps | [3,75] Tbps/ km$^2$ | [7,5] Tbps/ km$^2$ | Pedestrians |
| Dense urban | 300 Mbps | 50 Mbps | 750 Gbps/km$^2$ | 125 Gbps/km$^2$ | Pedestrians and users in vehicles (up to 60 km/h) |
| Broadcast-like services | Maximum 200 Mbps (per channel) | N/A or modest (e.g. 500 kbps per user) | N/A | N/A | Stationary users, pedestrians, and users in vehicles (up to 500 km/h) |
| High-speed train | 50 Mbps | 25 Mbps | 15 Gbps/train | 7,5 Gbps/train | Users in trains (up to 500 km/h) |
| High-speed vehicle | 50 Mbps | 25 Mbps | [100] Gbps/ km$^2$ | [50] Gbps/km$^2$ | Users in vehicles (up to 250 km/h) |
| Airplanes connectivity | 15 Mbps | 7,5 Mbps | 1,2 Gbps/plane | 600 Mbps/plane | Users in airplanes (up to 1,000 km/h) |

DL = downlink, UL = uplink.

### 10.6.1 Scenarios and KPIs

Different deployments of URLLC capabilities will depend on the 5G system being able to meet specific sets of KPIs with different values and ranges applicable for each attribute. A common yet flexible 5G approach to URLLC will enable the 5G system to meet the specific sets of KPIs needed in a given implementation. To provide clear and precise requirements for specific types of services, the corresponding KPI requirements are included in other specifications, such as for cyber-physical control applications in vertical domains [15], V2X [16], or rail communications [17].

## 10.7 Summary

The discussion in this chapter aimed to outline old and new criteria for designing a successful virtualized 5G system. Virtualized systems need a different approach. One needs to consider the coupling of hardware and software as well as incorporating the methods and development from computer science for software engineering, data processing, and computing concepts. The performance targets for 5G systems are challenging and add to the complexity of engineering virtualized networks. However, we believe that if the engineering team is conscious of the requirements and of the techniques needed for the solution, then the engineering of an advanced virtualized 5G can be successful.

# References


[1] Beck, K., et al. "Manifesto for Agile Software Development," 2001, http://agilemanifesto.org/.

[2] Ebert, C., G. Gallardo, J. Hernantes, and N. Serrano, "DevOps," *IEEE Software,* Vol. 33, No. 3, 2016, pp. 94–100.

[3] Salkin, H. M., and C. A. De Kluyver, "The Knapsack Problem: A Survey," *Naval Research Logistics Quarterly,* Vol. 22, No. 1, 1975, pp. 127–144.

[4] Andersen, David, et al., "Resilient overlay networks," *Proceedings of the eighteenth ACM symposium on Operating systems principles,* 2001.

[5] https://osm.etsi.org.

[6] https://www.cncf.io.

[7] https://kubernetes.io/de.

[8] Bromhead, B., "10 Advantages of Open Source for the Enterprise," https://opensource.com/article/17/8/enterprise-open-source-advantages?intcmp=701f2000000tjyaAAA&extIdCarryOver=true&sc_cid=701f2000001OH8CAAW.

[9] Bigelow, S., "What Are the Types of Requirements in Software Engineering?" https://www.techtarget.com/searchsoftwarequality/answer/What-are-requirements-types.

[10] Wikipeadia, Creative Commons (CC) licenses, https://en.wikipedia.org/wiki/Creative_Commons_license.

[11] Wikipedia, GNU General Public License, https://en.wikipedia.org/wiki/GNU_General_Public_Liscense.

[12] 3GPP/ETSS, Service Requirements for the 5G System, Version 16.14.0 Release 16, www.etsi.org, April 2021.

[13] ACM, ACM Computing Classification System; https://dl.acm.org/ccs.

[14] ITU-T, IMT Vision–Framework and Overall Objectives of the Future Development of IMT for 2020 and Beyond, itu.int.

[15] 3GPP TS 22.104, Service Requirements for Cyber-Physical Control Applications in Vertical Domains.

[16] 3GPP TS 22.186, Enhancement of 3GPP Support for V2X Scenarios.

[17] 3GPP TS 22.289, Mobile Communication System for Railways.

# 11

# Scaling Disaggregated vRANs

## 11.1 The Disaggregated vRAN

The deployment of 5G is expected to enable network densification along with the support of richer and more demanding applications. Consequently, the RAN—a key component of the cellular network infrastructure—is inevitably going to grow increasingly more complex. To tackle and manage this growing complexity, it is critical to automate the process of deploying, optimizing, and operating the RAN while also taking full advantage of newly available data-driven technologies to ultimately improve the end-user quality of experience. Many service providers have therefore opted for the disaggregation of the traditional monolithic RAN architecture and introduction of a RIC architecture in which the RAN control and data planes are decoupled. Hardware disaggregation requires network openness, which in turn attracts and facilitates the broadest development of applications, including AI-enabled applications that are specifically designed to automate operations and optimize the RAN while accounting for its continuous evolution toward more high-performing deployments.

### 11.1.1 RAN Disaggregation

The 5G cellular network has been standardized to meet diverse demands that are classified into three broad categories: eMBB, URLLC, and mMTC. Due to these broad objectives the evolution toward and beyond 5G calls for an architectural transformation to support service heterogeneity, coordination of multiconnectivity technologies, and on-demand service deployment. The cost involved in the deployment, optimization, and operation of the RAN components generally accounts for approximately 70% of the total network cost [1]. Not surprisingly, the RAN is the most targeted candidate by operators for decreasing the overall network expenditure. Traditionally, the RAN base station (also known as gNB for 5G, and eNB for 4G) consists of a BBU and many radio units (RUs). The BBU implements NFs across the protocol stack, whereas the RUs handle radio wave transmissions to and from the user equipment. RAN disaggregation aims to split the BBU functionalities into two entities: the DU and the CU. The CP and UP separation is implemented using SDN and NFV methodologies. On the one hand, the SDN approach decouples the CP NFs from both the RAN and data plane NFs, running them inside a centralized controller that interacts with the disaggregated hardware via standard interfaces. Additionally, the NFV approach requires NFs, which conventionally run on specialized RAN hardware, to be redesigned to run in VMs that are instantiated in either cloud infrastructure and/or dedicated COTS hardware.

Figure 11.1 provides a visual comparison between the conventional RAN and disaggregated RAN. The conventional RAN consists of a monolithic system where the radio antenna units and the BBU are typically situated at the cell tower. In the conventional RAN, the evolved packet core is implemented as a single system running in the centralized cloud. In contrast, the disaggregated RAN consists of the radio antenna unit situated at the cell tower followed by the 5G base station DU and CU, both running in the edge cloud. The CU

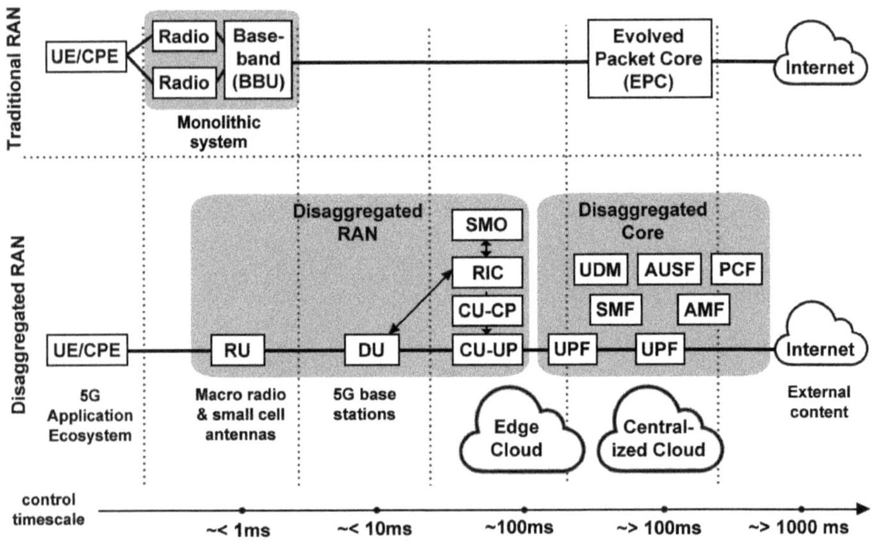


**Figure 11.1** Conventional vs disaggregated RAN.

implements the CP and UP as separate entities. The packet core is disaggregated and comprises multiple modules (e.g., AMF and UDM).

RAN disaggregation offers many potential benefits to network operators, including reduced capital investment, lower operating expenses, architectures built around use cases rather than around vendors, and a more robust supply chain [2, 3]. RAN disaggregation must be coupled with an intelligent controller that handles the 5G RAN deployment, optimization, and management while guaranteeing the support of demanding service requirements such as network slicing, high bandwidth connectivity, and low-latency applications.

## 11.2 RAN Intelligent Controller Overview

RIC is a software-defined component of the Open-Radio Access Network (O-RAN) [4] architecture devised to bring intelligence, programmability, and extensibility capabilities to the 5G RAN. At a high level, RIC is an SDN-based component that performs selected radio resource management (RRM) functions that were traditionally managed in the monolithic CU-CP solution. RIC implements AI and ML algorithms to efficiently automate RAN operations and deliver unprecedented flexibility to multiple use cases. As shown in Figure 11.2 and defined in the O-RAN architecture, the RIC architecture consists of two radio intelligent controller modules (i.e., non-RT RIC and near-RT RIC), which together aim to enhance the traditional RAN functions with embedded intelligence. Figure 11.2 shows the service management and orchestration (SMO) component that hosts the non-RT RIC to provide policies/intent and enrichment information concerning RRM to the near-RT RIC.

### 11.2.1 Interfaces

The 3GPP specifications define two standard interfaces to support RAN disaggregation: (1) F1 between the CU unit and DUs, and (2) E1 between the monolithic CU-CP unit and its CU-UPs. The O-RAN Alliance defines additional interfaces: (1) E2 connecting near-RT RIC with underlying RAN nodes, (2) A1 interfacing near-RT RIC with SMO, and (3) O1 connecting SMO with near-RT RIC and underlying RAN nodes. Here is a summary of these interfaces:

*F1 interface:* The interface between CU unit of a gNB and its DUs.

*E1 interface:* The interface between the CU-CP unit of a gNB and its CU-UPs.

*E2 interface:* The E2 interface publishes fine-grained per-UE data from the RAN node to near-RT RIC for consumption by RRM microservices

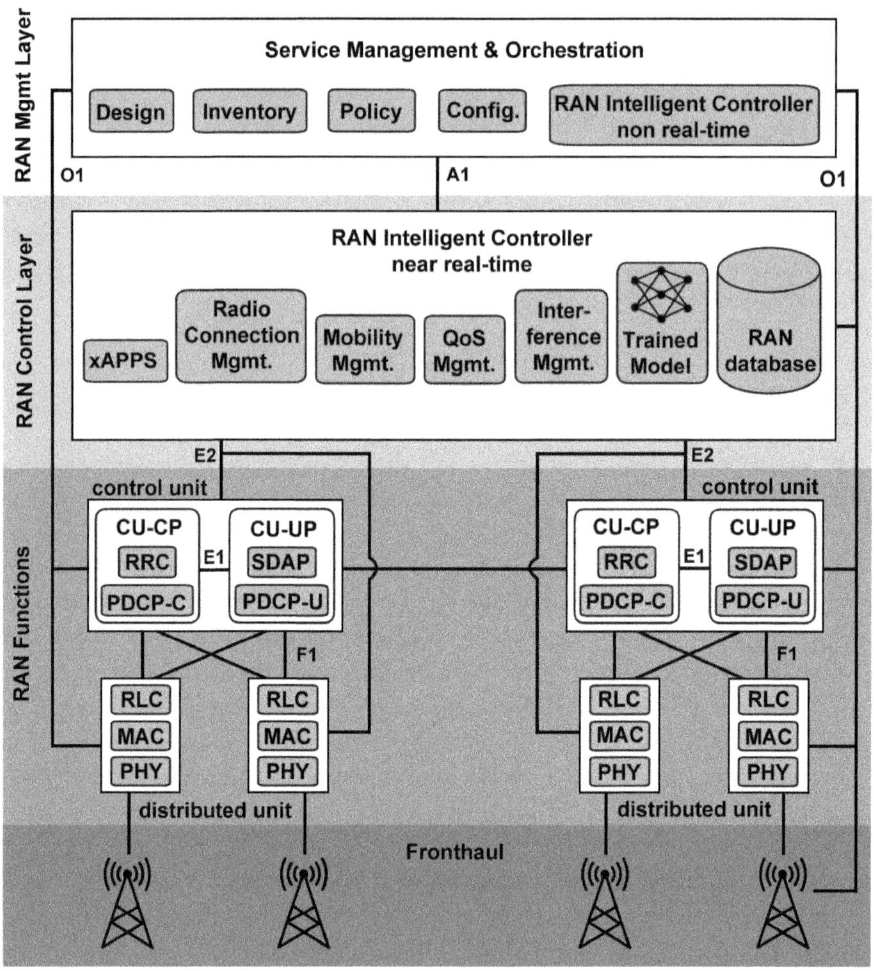


**Figure 11.2** Overview of the RIC functional architecture.

that produce control actions and/or policies directed toward the CU or DU.

*A1 interface:* The northbound A1 interface interconnects non-RT RIC (inside SMO) and near-RT RIC. AI/ML-based models concerning mobility management, spectrum resource, QoS management, load balancing, interference coordination, and RAN slicing can generate enhanced policy guidance to near-RT RIC via A1.

*O1 interface:* Northbound operations, administration, and management interface of RIC. It is responsible for onboarding, deployment, activation, life cycle, and computational resource management of near-RT RIC for performance guarantees.

### 11.2.2 RIC Design Principles and Components

In the conventional RAN, CP processing occurs inside the CU-CP. In the disaggregated design, CP processing follows an event-driven approach involving near-RT RIC, that is, when a RAN event occurs, the relevant CP processing in the CU-CP is temporarily suspended (using E2 interface INSERT/CONTROL mechanism), delegating the next step to be performed by RIC using one of its microservices (xApps). The output generated by the ML-enabled xApp algorithms is subsequently delivered back to the CU-CP in the form of a policy (e.g., handoff, admission control, and bearer reconfiguration) for completion of the CP processing.

The RIC key components are:

*RAN database:* A RAN network information database that stores the status information of the RAN including topology and metrics. The RAN nodes can send their state information following a triggering event or at regular periodic intervals.

*E2 termination:* This provides an interface between near-RT RIC and the RAN via Stream Controlled Transmission Protocol (SCTP).

*E2 manager:* This provides an API for managing the connection (i.e., connect, disconnect, reset) between near-RT RIC and the RAN elements.

*xApps:* Software applications that run on near-RT RIC to perform RRM tasks under the guidance of ML/AI. The xAPPs and the RAN database are deployed as microservices in VM clusters on COTS platforms.

*xApp manager:* This provides APIs for managing the lifecycle (deployment and removal) and configuration of xApps.

*A1 mediator:* This acts as the interface between SMO and near-RT RIC to provide policy guidance to the xApps.

### 11.2.3 Policy Guidance

The RIC xApps make optimized RRM decisions subject to the policy guidance provided by SMO to near-RT RIC over the A1 interface. Some examples of RAN policy-guidance include:

- KPI optimization and service intent/objectives (e.g., cell-specific throughput, spectrum utilization, and UE-specific guaranteed bit rate);
- Providing additional information (A1-EI) from outside the RAN ecosystem to RIC (e.g., weather information and vehicular traffic congestion data) for use by ML-based xApps as input features;

- Prioritizing or blacklisting cells and frequency bands for carrier aggregation, traffic offload, load balancing, and so on.

### 11.2.4 ML/AI Role in the RIC

With the recent advances in AI and ML techniques, automation in cellular networks has gained momentum using AI/ML for optimizing video flows and traffic prediction, improving network energy efficiency, and resource allocation. The role of ML/AI is to improve the RRM and varies from RAN submodule to submodule depending on the control/timescale requirements as shown in Figure 11.1.

RAN KPI statistics reported by the RAN nodes to SMO over the O1 interface are used as training datasets for the AI/ML models in non-RT RIC. SMO uses these models to improve the accuracy of decisions on both policies and configuration of KPI objectives. These decisions are further communicated near-RT RIC over A1.

Due to the near-RT interaction between the CU and near-RT RIC, the ML-based microservices hosted in near-RT RIC use hybrid models, comprising a mix of both offline and online ML. Online ML models (e.g., deep/recurrent neural networks, reinforcement learning) can achieve lower control loop latency since they are typically processed on a single stream of incoming RAN data delivered to near-RT RIC. Due to the resulting poor accuracy in generating optimized RRM decisions based on incoming RAN data, additional offline ML models are used that make use of historical information previously stored in the RIC database. Typically, short-lived inferences last milliseconds, whereas longer time-scale analysis is continuously performed by non-RT RIC. Feedback on the precision and accuracy of the predictions is provided to non-RT RIC via the O1 interface and is applied to fine-tune the ML models and guide the operation of the overall application toward a chosen objective. If any ML model in near-RT RIC misbehaves or exhibits degraded performance, non-RT RIC may instruct near-RT RIC to terminate that ML model or switch to an alternative model. This approach yields improved RRM decisions made by RIC in near-real-time.

The stringent latency requirements imposed on the communications protocols between DUs and RUs may prevent a correct use of the request-response approach between DUs and RIC. However, asynchronous policy directives can still be provided by the RIC in this case. An example of a policy directive from RIC to each DU is the choice of scheduling algorithm (e.g., proportional-fair versus round-robin) that each DU must use, subject to extant RAN conditions. This AI-based policy directive can be leveraged by DUs to improve RRM performance in real-time.

## 11.3 Security Challenges

Wireless communication systems have been prone to security vulnerabilities from their inception. In the first generation (1G) wireless networks, mobile phones and wireless channels were targeted to achieve illegal cloning and masquerading. In the second generation (2G) wireless networks, message spamming became a common form of pervasive attacks and injection of false information leading to broadcasting of unwanted marketing advertisements. In the third generation (3G) wireless networks, IP technology became part of the mobile architecture, unfortunately bringing along the typical internet security vulnerabilities and challenges. Fourth generation (4G) mobile networks paved the way to the proliferation of smart devices and multimedia traffic, which in turn created a more complicated and dynamic threat landscape. 5G provides increased broadband access, entertains higher user mobility, and enables massive device connectivity with the help of technology enablers such as cloud computing, software-defined networking, and network function virtualization. There are pressing security issues for these advanced technologies which are highlighted in the remainder of this chapter.

### 11.3.1 Key Security Threats

Figure 11.3 highlights the potential sources of attacks in the 5G network. These attacks could come from many sources such as end devices, untrusted networks, roaming networks, the internet, and application service providers. They are categorized as loss of availability, loss of confidentiality, loss of integrity, loss of control, loss of integrity due to malicious insiders, and theft of service, respectively. The attacker(s) can launch any of these types of attacks by various means and consequently pose serious threats to the network assets. For example, an attacker can launch a DoS attack and make the network unavailable by flooding a specific network interface or crashing a network element. Similarly, an attacker can launch man-in-the-middle attacks to modify the traffic by way of eavesdropping. The basic challenges in 5G—as highlighted by the Next Generation Mobile Network (NGMN) organization and broadly discussed in the literature [5, 6]—are as follows:

*Flooding an interface:* Attackers flood an interface resulting in a DoS condition (e.g., multiple authentication failures on N1, N2 interface) (loss of availability);

*Crashing a network element:* Attackers crash a network element by sending malformed data packets (loss of availability);

*Eavesdropping:* Attackers eavesdrop on sensitive data transmitted on control and bearer planes (loss of confidentiality);

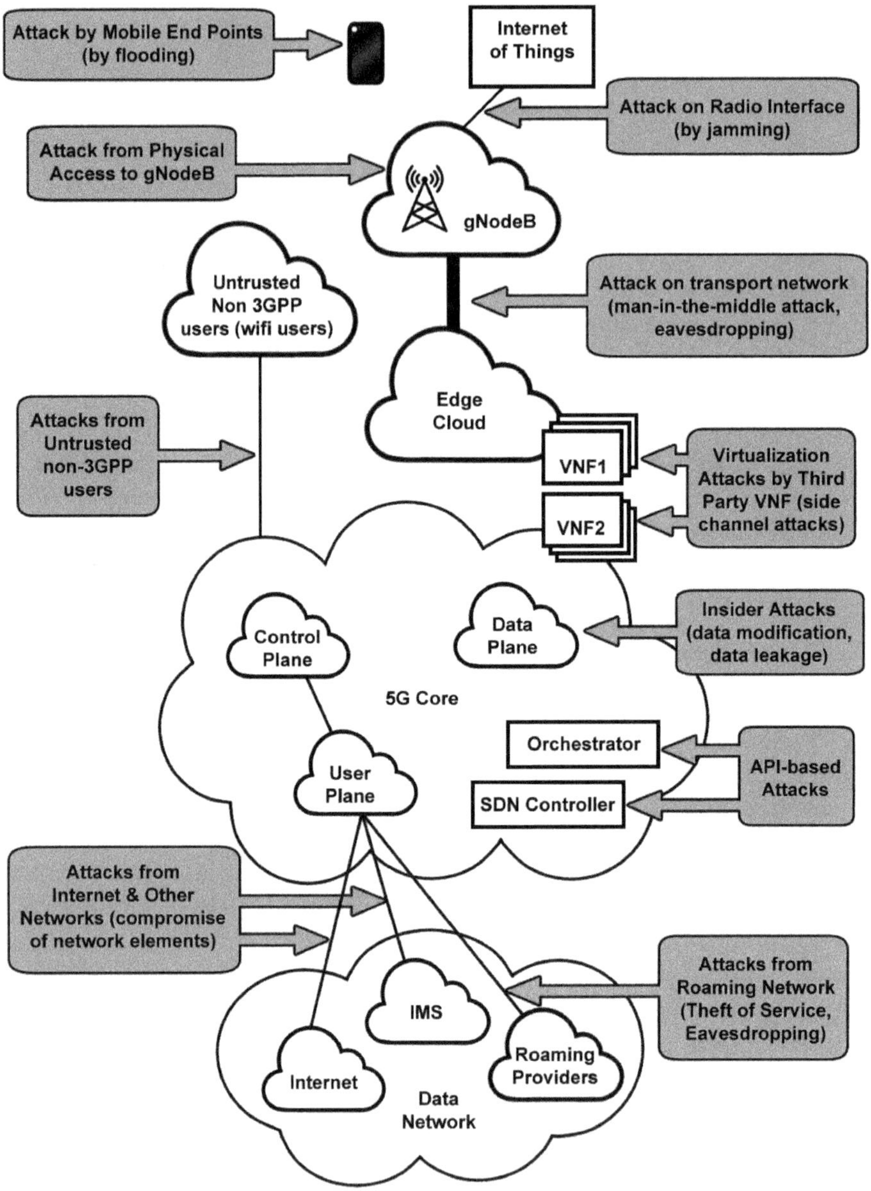


**Figure 11.3** Key security threats in 5G.

*Data leakage:* Unauthorized access to sensitive data on the server profile, for example, Unified Data Repository (UDR) and Unrestricted Data Storage Function (UDSF) (loss of confidentiality);

*Traffic modification:* Attackers modify information during transit in user plane N3, for example, Session Initiation Protocol (SIP) header modification and Real-time Transport Protocol (RTP) spoofing (loss of integrity);

*Data/configuration modification:* Attackers modify data on network element (loss of integrity);

*Network control:* Attackers control the network via protocol or implementation flaws (loss of control);

*Compromise of network element:* Attackers compromise network elements via a management interface (loss of control);

*Insider threats:* Insiders make data modifications on network elements or make unauthorized changes to network element configuration (malicious insider);

*Fraud or configuration modification:* Attackers exploit a flaw to use services without being charged (theft of service).

### 11.3.2 Key Security Pillars

Figure 11.4 illustrates the key security pillars in 5G that are commonly identified in the literature. Identifying and distinguishing between these different security pillars can help us formulate corresponding and suitable mitigation

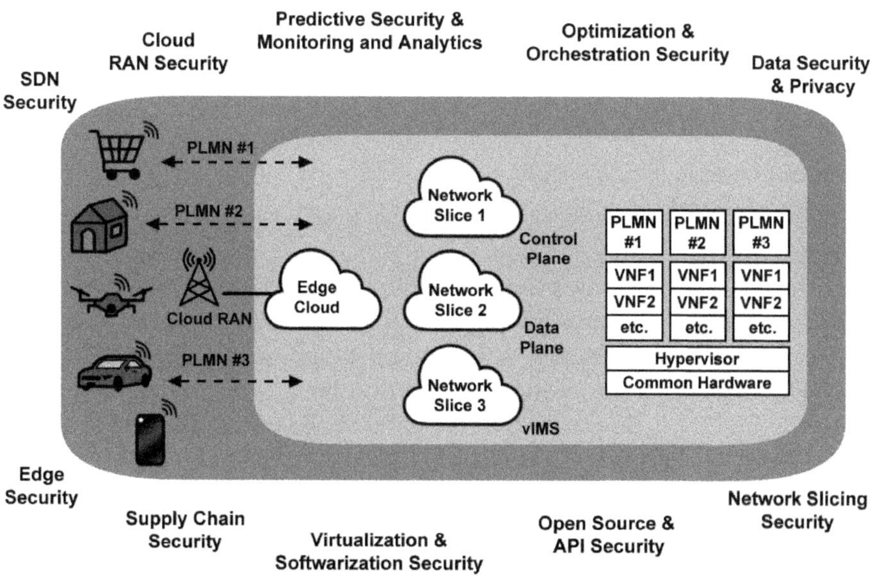


**Figure 11.4** Key security pillars in 5G. (*From:* [6].)

techniques. The vulnerabilities associated with these security pillars and the related possible security solutions are discussed in the next section.

#### 11.3.2.1 Virtualization and Softwarization Security

Since the successful introduction of virtualization, hypervisors and containers have gained significant traction. If on the one hand these technologies allow multiple tenants and virtual network functions to reside on the same physical hardware, on the other they increase the systems' attack surface to threats such as data exfiltration, resource starvation, and side-channel attacks. Mitigation techniques that can be applied to such scenarios include hypervisor introspection schemes and hypervisor hardening. These techniques can protect the hypervisor's code and data from unauthorized modification and can guard against misconfigurations.

#### 11.3.2.2 Open-Source and API Security

There are various open-source activities targeting 5G, including the Open Networking Foundation (ONF), OpenDaylight (ODL), Open Network Operating System (ONOS), Open vSwitch (OVS), and the Linux Foundation. The operator and vendor communities are collaborating to develop open-source solutions that are scalable and reliable enough to be deployed. While open-source efforts offer various advantages such as flexibility and agility, faster time to market, and cost-effectiveness, they also present challenges and shortcomings, such as limited level of support, intellectual property concerns, lack of documentation and GUIs, and lack of customization needed to address specific use cases. These challenges and shortcomings give rise to security concerns that need to be addressed by the open-source community. The pros and cons of open-source security challenges are detailed in Chapters 9 and 10.

#### 11.3.2.3 Network Slicing Security

Network slicing is the ability to partition network resources to efficiently support different types of applications on the same physical network. Proper security controls must be implemented to ensure robust isolation of resource slices from one another and achieve a trusted virtualization infrastructure. Such security controls include slice categorization and adequate provisioning of resources. In addition, strong security mechanisms must be in place to limit and secure information flow within each slice. For instance, security as a service (SECaaS) is a multitenant, scalable, virtual, and customized service that is offered with usage/subscription-based pricing. Integrating SECaaS to network slicing [1] can prevent and mitigate many threats such as side-channel attacks across slices and DoS attacks via virtual resource depletion.

#### 11.3.2.4 SDN Security

SDN centralizes network control and enables programmability of communication networks. These two features, however, create opportunities for cracking and hacking the network. For example, a centralized SDN controller periodically modifies flow rules in the data path and can therefore be easily identified by a DoS attacker who can expose critical APIs to unintended software and take the whole network down [7]. The centralization of network control can also make the controller a bottleneck for the whole network due to saturation attacks. Since most network functions can be implemented as SDN applications, if granted access malicious applications can spread havoc across a network [7]. Therefore, detection and mitigation of such attacks represent one of the utmost important aspects of security. ML can be exploited for this purpose [8]. However, ML is also vulnerable to adversarial attacks. Therefore, a secure system is needed to prevent malicious data from being injected into the ML training phase. Several solutions for robust classification and anomaly detection have been proposed in the literature [9].

#### 11.3.2.5 Cloud RAN Security

One of the main differences between conventional RAN and C-RAN is that cloud computing is applied in the latter. Thus, it is critical to consider cloud-computing security problems. 5G networks will enable many more devices (i.e., IoT) to be connected to the RAN through shared access. Therefore, a multitude of infected IoT devices attempting to gain access may result in shared resource overload, VM/Guest OS manipulation, and data exfiltration. In [10], the authors discuss the threats and security challenges of the cloud system, and define the basic requirements for building a secure and trustworthy cloud system, which are: (1) outsourcing security: the cloud provider must be trustworthy by providing trust and privacy protection, and they should ensure the confidentiality and integrity of the outsourced data; (2) multitenancy security: the shared cloud platform must ensure the security of resource allocation in a virtualized environment; (3) massive data and intense computation security: it is necessary to design new secure strategies and protocols to achieve massive data and intense computation.

One survey on C-RAN security presents nine security threats in cloud computing [11]. For C-RAN, the following security threats should be carefully considered: data loss and leakage, shared technology issues, abuse and nefarious use of cloud services, and DoS attacks. Deploying both DoS detection and dynamic service chaining can act as mitigation steps. Therefore, Cloud RAN in 5G networks must have embedded DoS detection and related mitigation functions.

#### 11.3.2.6 Edge Computing Security

MEC is a recently introduced approach in the 5G ecosystem for the provision of cloud computing capabilities at the edge of the network to support high-bandwidth, low-latency end-user applications. MEC is in the logical proximity of base stations enabling authorized third parties to offer close-by processing and storage capabilities to subscribers of the 5G network. MEC security threats relate to components that are located at the edge of the network, which become attractive targets for cyberattacks. Most of these attacks fall under the categories of man-in-the-middle attacks and eavesdropping. The edge modules host security controls such as authentication, authorization, and real-time attack detection to provide security controls to 5G use cases, which may lead to additional system vulnerability. In this case, strong layered security controls must be implemented at the edge to achieve adequate protection.

#### 11.3.2.7 Supply Chain Security

The deployment of modular commodity hardware and software components introduces a multitude of additional security risks. Threats include compromising products through concealed hardware, malicious software, and software flaws. Threats also include implementation of uncontrolled software updates, manipulation of functionalities, inclusion of functions to bypass audit mechanisms, backdoors, and undocumented testing features that may propagate into the production version. These security threats require solutions that operate at multiple levels. Computationally feasible trust platforms such as blockchain may help establish some security controls over commodity hardware and integrated software.

#### 11.3.2.8 Data Security and Privacy

Being a complex ecosystem, the 5G network will be used by many types of players, including infrastructure providers, mobile communication network operators, and virtual operators. The storage, transmission, and processing of user data within such a complex network ecosystem is likely to lead to user data being duplicated and scattered across several network elements/sites. Unnecessary data duplication leads to additional security threats including multiple network entry points—such as user equipment (mobile devices and IoT)—and operation and management interfaces. Authentication threat includes the theft of user credentials, brute force hacking of user accounts, password cracking, masking of user identity, and impairment of IoT authentication. From a security perspective, several cases should be considered, including classification and proper protection for at-rest and in-transit data. Data privacy and regulatory requirements need to be considered when designing/configuring the system to ensure that only strictly necessary data is collected and stored, while at the same time being compliant with applicable regulations.

#### 11.3.2.9 Optimization and Orchestration Security

Resource orchestrator is one of the most important components of the 5G infrastructure. It is responsible for the configuration and management of all significant 5G components, including NFV, VNF management, and virtualized infrastructure management (VIM). In an SDN/NFV environment, an orchestrator could provision VNFs based on the network condition and intelligence. This built-in orchestration flexibility introduces potential vulnerabilities, whereby an attacker may use legitimate access to the orchestrator to manipulate its configuration to run a compromised VNF.

#### 11.3.2.10 Predictive Security Monitoring and Analytics

While it may be effective to detect cyberattacks quickly and be able to mitigate them in a timely manner, stopping the attacks altogether by taking proactive measures is highly desirable. This can be achieved by applying AI/ML techniques to detect anomalous behavior, creating behavior analytics of bad actors through traffic analysis and deep packet inspection, and recognizing patterns of recorded past attacks. This approach can improve detection and mitigation of the zero-day attack problem, which is the ability of hackers to exploit system vulnerability before software developers can find a fix.

In this chapter, so far, we have highlighted the main security challenges of the 5G network. Operators, vendors, research labs, and regulators need to work together to define a security ecosystem for future networks based on continuously evolving technologies and rapidly evolving threat landscapes.

## 11.4 5G Resiliency

Service availability is one of the most important characteristics of mobile networks. Generally, for all mission and business-critical services, five nines (99.999 percent) KPIs are the norm. Ensuring five nines availability in the network is not a trivial task. Whenever a network node, site, or cell goes down there are significant consequences for both the cloud service provider and the vendors involved. In addition to contractual SLAs between the cloud service provider and the vendor being violated, the ensuing economic negative impact of the unavailable mobile network service and the negative publicity for the cloud service provider must be considered. In summary, the five nines availability of mobile networks is and will become increasingly important.

### 11.4.1 Network Resiliency

The inherent geographically distributed C-RAN architecture requires a high-speed and low-latency transport network. Notwithstanding right-of-way access, optical fiber cables, and dense wavelength division multiplexing (DWDM) represent the most desirable solution for the RAN due to their abundance of data transmission capacity. Proprietary or open optical network solutions are now offering physical layer SDN programmability that can be readily leveraged to achieve highly reliable transport connectivity. One such scenario is explained in detail below.

The principles of SDN are extended in a programmable optical network orchestrator to control and coordinate the use of the equipment at both the Ethernet (upper) and DWDM (lower) layers. A programmable optical network with orchestrator support enables on-demand data flow provisioning and coordination of fault handling at both the Ethernet and DWDM layers. The orchestrator discovers the network topology at both layers, computes two disjoint paths between the endpoints of incoming flow requests, and executes recovery procedures in the event of a network failure (e.g., a fiber cut). By jointly controlling the resources at the two layers, the orchestrator can efficiently combine restoration and protection mechanisms together, thus achieving a highly reliable multilayer network. Restoration mechanisms are implemented at the optical layer (i.e., when an optical circuit fails, a restoration circuit is automatically established by the orchestrator). Protection mechanisms are implemented at the Ethernet layer using fast failover table protection (i.e., two disjoint flows are established ahead of the failure, thus ensuring that at least one flow is readily available and functional in the presence of any single network failure). The fronthaul and backhaul link protection mechanisms will vary based on the functionality split and latency constraints. One such implementation of these protection mechanisms is described in [12].

### 11.4.2 VNF Resiliency

While transitioning from 4G to 5G deployments, network providers are leveraging both SDN and NFV) concepts to implement cost-effective and scalable cloud-native 5G network solutions. Realizing that a monolithic implementation of these VNFs usually requires a long time in both deployment and upgrade cycles [13], an alternative cloud-native approach based on containerized network functions (CNFs) is gaining traction as it is believed to accelerate software implementation of network functions.

Despite their potential benefits, software-based network functions are more vulnerable to failures than traditional middleboxes. CNF platforms often require additional virtual layers between CNFs and the underlying hardware. Any misconfiguration of these intermediate layers may lead to CNFs failures. In addition, the reliability of dedicated hardware is inherently superior to that of commodity servers. Therefore, guaranteeing the high availability of CNFs—which is essential to providing reliable network services—presents new challenges. Deploying backups is a well-established and robust method to improve the availability of CNFs. However, it comes with the extra cost of fully duplicated resources.

Realizing the importance of timely offering CNF, the computing platform must include features such as live migration, snapshot, and rebirth to ensure that the SLA requirements are met in terms of security, reliability, and total cost of ownership. Live migration is the process of migrating CNFs from one node to another while guaranteeing zero or minimal impact on the connectivity services offered to mobile users. Being able to live migrate CNFs of the virtual RAN offers several significant advantages, including (1) achieving load balancing in the compute nodes [14] by timely redistributing VNFs to sparsely loaded servers, (2) effectively performing regular maintenance—such as upgrading OS versions and changing network configurations—and (3) more efficiently handling fault management. These features may also prove to be of the essence to cope with the highly fluctuating mobile traffic volumes that are expected in real-time networks. With that said, dynamic scaling of telco-supported services is quite different from general cloud applications and requires additional study.

More recently, methods have been proposed to specifically address failures when 5G functions are virtualized. In [15], a two-step resiliency scheme is proposed to overcome soft failures of the optical circuits (lightpath) transporting fronthaul traffic. When resources available along the backup lightpath are insufficient, the radio functional split is dynamically reconfigured to reduce fronthaul bandwidth requirements. In [16], the authors evaluated the container-based live migration of video applications to meet the multiaccess edge computing requirements during radio handover. The CNF migration of virtualized core network using checkpoint/restore in userspace (CRIU) is discussed in [17].

### 11.4.3 Dynamic Rerouting with Live Migration Support

Fronthaul, midhaul, and backhaul networks rely on Ethernet-over-DWDM transport network technologies in which resiliency to failures is achieved through dynamic rerouting [18]. The implemented fault tolerance schemes must also consider the effects that any network element failure has on the application's QoS guarantees. For example, a longer secondary path that replaces a shorter failing primary path introduces additional signal propagation delay that may violate the application minimum QoS requirements. Advanced recovery mechanisms that combine connection rerouting and VNF live-migration may become necessary.

The scenario in Figure 11.5 illustrates how dynamic network rerouting and VNF live-migration may be concurrently applied. In a typical cloud architecture, the resource orchestrator interacts with the VIM. The NFV infrastructure consists of compute, storage, and network resources that are managed by VIMs. The VNF manager instantiates and monitors VNF instances. When

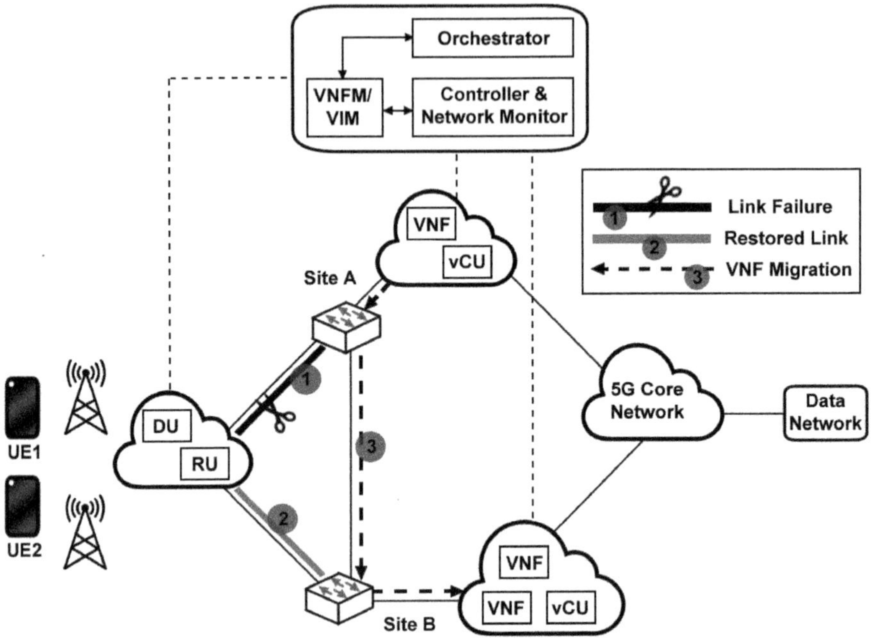


**Figure 11.5** Dynamic rerouting with live migration support.

a network path/link failure occurs, traffic rerouting is initiated by the orchestrator to circumvent the failing network element. Concurrently, as shown in Figure 11.5 by the dashed arrows, the virtual CU component is live migrated from Site A to Site B to meet the application's network latency requirements after rerouting.

To achieve five nines of availability in 5G mobile network, AI is expected to play a central role by predicting network element failures ahead of time and preparing appropriate proactive mitigation procedures. AI must be able to predict a node's future downtime based on the network history. Unlike conventional RANs in which resources are statically allocated, C-RAN solutions are expected to leverage AI predictions to proactively and dynamically reserve resources to account for network slicing, application requirements, and potential element failures in future 5G mobile networks.

## References

[1] Pla, L. F., N. Shashidhar, and C. Varol, "On-Premises Versus SECaaS Security Models," in *Proceedings of the 8th International Symposium on Digital Forensics and Security (ISDFS)*, Beirut, Lebanon, June 1–2, 2020, pp. 1–6.

[2] Gkatzios, N., M. Anastasopoulos, A. Tzanakaki and D. Simeonidou, "Compute Resource Disaggregation: An Enabler for Efficient 5G RAN Softwarisation," *2018 European Conference on Networks and Communications (EuCNC),* 2018, pp. 1–5, doi: 10.1109/EuCNC.2018.8443270.

[3] Balasubramanian, B., et al., "RIC: A RAN Intelligent Controller Platform for AI-Enabled Cellular Networks," *IEEE Internet Computing,* Vol. 25, No. 2, March–April 2021, pp. 7–17, doi: 10.1109/MIC.2021.3062487.

[4] Polese, M., L. Bonati, S. D'Oro, S. Basagni, and T. Melodia, "Understanding O-RAN: Architecture, Interfaces, Algorithms, Security, and Research Challenges," in arXiv 2022, https://doi.org/10.48550/arXiv.2202.01032.

[5] Ahmad, I., T. Kumar, M. Liyanage, J. Okwuibe, M. Ylianttila, and A. Gurtov, "Overview of 5G Security Challenges and Solutions," *IEEE Communications Standards Magazine,* Vol. 2, No. 1, March 2018, pp. 36–43, doi: 10.1109/MCOMSTD.2018.1700063.

[6] Dutta, A., and E. Hammad, "5G Security Challenges and Opportunities: A System Approach," in *2020 IEEE 3rd 5G World Forum (5GWF),* 2020, pp. 109–114, doi: 10.1109/5GWF49715.2020.9221122.

[7] Mathebula, I., B. Isong, N. Gasela and A. M. Abu-Mahfouz, "Analysis of SDN-Based Security Challenges and Solution Approaches for SDWSN Usage," in *2019 IEEE 28th International Symposium on Industrial Electronics (ISIE),* 2019, pp. 1288–1293, doi: 10.1109/ISIE.2019.8781268.

[8] Benzaïd, C., M. Boukhalfa, and T. Taleb, "Robust Self-Protection Against Application-Layer (D)DoS Attacks in SDN Environment," in *Proceedings of the 2020 IEEE Wireless Communications and Networking Conference (WCNC),* May 2020, pp. 1–6.

[9] Qin, Y., J. Wei, and W. Yang, "Deep Learning Based Anomaly Detection Scheme in Software-Defined Networking," in *2019 20th Asia-Pacific Network Operations and Management Symposium (APNOMS),* 2019, pp. 1–4, doi: 10.23919/APNOMS.2019.8892873.

[10] Xiao, Z., and Y. Xiao, "Security and Privacy in Cloud Computing," *IEEE Communications Surveys & Tutorials,* Vol. 15, No. 2, Second Quarter 2013, pp. 843–859, doi: 10.1109/SURV.2012.060912.00182.

[11] Tian, F., P. Zhang, and Z. Yan, "A Survey on C-RAN Security," *IEEE Access,* Vol. 5, pp. 13372–13386, 2017, doi: 10.1109/ACCESS.2017.2717852.

[12] Ramanathan, S., M. Tacca, M. Razo, et al., "A Programmable Optical Network Testbed in Support of C-RAN: A Reliability Study," *Photonic Network Communications,* Vol. 37, No. 3, pp. 311–321.

[13] JayaKumar, M., "Why Use Containers and Cloud-Native Functions Anyway?" https://www.intel.com/content/www/us/en/communications/why-containers-and-cloud-native-functions-paper.html.

[14] Yu, J. F. R., S. Wang, T. Huang, Z. Liu, and Y. Liu, "Load Balancing in Data Center Networks: A Survey," *IEEE Communications Surveys Tutorials,* Vol. 20, No. 3, 2018, pp. 2324–2352.

[15] K. Kondepu, A. Sgambelluri, N. Sambo, F. Giannone, P. Castoldi, and L. Valcarenghi, "Orchestrating lightpath Recovery and Flexible Functional Split to Preserve Virtualized RAN Cconnectivity," *Journal of Optical Communications and Networking,* Vol. 10, No. 11, November 2018, pp. 843–851.

[16] Addad, R. A., D. L. C. Dutra, M. Bagaa, T. Taleb, and H. Flinck, "Fast Service Migration in 5G Trends and Scenarios," *IEEE Network,* Vol. 34, No. 2, 2020, pp. 92–98.

[17] Ramanathan, S., K. Kondepu, M. Razo, et al., "Live Migration of Virtual Machine and Container Based Mobile Core Network Components: A Comprehensive Study," *IEEE Access,* Vol. 9, 2021, pp. 105082–105100.

[18] Ramanathan, S., K. Kondepu, T. Zhang, et al., "Orchestrating Virtualized Core Network Migration in OpenROADM SDN-Enabled Network," in *2021 International Conference on Optical Network Design and Modeling (ONDM),* 2021, pp. 1–6.

# Part III
# Future Developments in the Mobile Network

# 12

# Private 5G Networks and the Edge

## 12.1 The Privatization of the Network with p5G

Private 5G (p5G) networks are nonpublic 5G networks and are an important option in the design of 5G systems. Some press and authors may also use terms such as P5G, private networking or networks, private LTE (pLTE), or a variety of other nomenclatures. p5G systems are targeted at enabling organizations to apply 5G technologies on their own premises and enabling the organizations to maintain full or partial control on these network systems. p5G systems can be implemented either as an independent 5G network or in conjunction with the public network through a business partnership with a carrier.

In this chapter, we will outline the technologies and deployment scenarios for p5G networks. The 5G Alliance for Connected Industries and Automation (5GACIA), which is a major forum for discussing applicability of 5G technology in connected industries [1], is a basis for this discussion. In addition, we discuss MEC since it extends the notion of a private network to private service at the edge. The term "edge" is heavily overloaded, by the operators, by industry, by the ecosystem, and nearly every other entity. Our use of edge simply implies it is not in a datacenter far from the original source or UE. MEC permits the implantation and mobility of services for private 5G networks where traffic may be shunted off the network near the source, while other traffic may be steered to systems deeper in the network. We outline the implications of MEC on the evolving private 5G network technologies. Finally, the chapter details some of business and challenges facing this category of 5G networks.

### 12.1.1 Usage Scenario and Objectives

Private 5G networks today are primarily used in communication tasks within factories, manufacturing, and industrial processes. The 5G standards found in ETSI and 3GPP pay particular attention to the use cases for IoT, in supporting URLLC and mMTC.

The standards for private 5G systems also address a vertical scenario in cyber-physical systems [2]. However, nomadic scenarios for logistics, mobility, or rescue [3] are also coming increasingly into focus. One reason is that the business needs of various enterprise users may have additional security requirements, including those required by regulation such as for Health Insurance Portability and Accountability Act (HIPAA) compliance (HIPAA is term found most commonly in medical records laws and regulations, but it may go by other terms in different jurisdictions), government security concerns, or business-driven needs (the European H2020 project "Bonseyes" calls this "keep my private data private," www.bonseyes.com).

p5G systems combine the advantages of both public and nonpublic 5G networks. This control over the network and ability to directly address other business needs are an appealing option across various industries, businesses, utilities, and the public sector. Moreover, they are considered as a promising accelerator for Industry 4.0 [4] (i.e., for factories and manufacturing processes that rely on the exchange of huge amounts of data and process information). The combination of these business drivers, technical capabilities, and the ability to now deploy at costs not previously possible, private 5G has recently gained significant research attention from both academia and industry.

Private 5G networks and this concept of private networking strives for unified connectivity, high quality of service, and, particularly, customized security and privacy within a dedicated coverage area. The need and the mechanisms for network and traffic isolation from other networks are an important design feature in private 5G. In addition, the requirements for operational safety (i.e., for tight responsibility for operations and maintenance (O&M) goes hand in hand with the security in order to assure mission critical functions and data protection).

Complementing the concept of private 5G networks is the notion of mobile edge computing. This concept has now materialized as an architecture and standards for future network by ETSI's initiative on MEC [5]. MEC provides an IT service environment and cloud-computing capabilities at the edge of the mobile network within the RAN and in close proximity to mobile end device or end connection. The aim is to reduce latency in the data plane, ensure highly efficient network operation and service delivery, and to offer an improved user experience. MEC focuses particularly on applications to be hosted in a multivendor and multiaccess edge environment.

### 12.1.2 Service Objectives and Attributes for Private 5G

Combining the above outlined scenarios, one can derive the service objectives and attributes for private 5G networks as detailed below that names the attribute in ordinary networks and provides an interpretation of the attribute in p5G systems.

#### Device Connectivity

Device connectivity describes the ability of the private 5G devices to connect to the private 5G network, but also to the public 5G system, and use their services.

A device might switch between internal services (i.e., private ones) and external services (i.e., provided by a public 5G system).

The feature includes the capability to provision and use services if a device moves to another geographic location. The public network can provide connectivity as well as it extends the private network to other geographical locations. In addition, the public 5G network can be used to provide access to public communication services while the device is still connected to the private systems.

#### Quality of Service

The quality-of-service attribute in private 5G systems is similar to public ones:

- Latency: maximum permissible end-to-end delay, may range of stringent (e.g., 1 ms) to modest values (e.g., 100 ms).
- Availability: ratio of time a service satisfies a specified QoS; ranges from stringent (e.g., 99.999999%) to modest values (e.g., 99.9%).

Many private 5G networks refined and augmented the above's base values to:

- Ultrareliable low latency: supporting near zero-latency, real-time needs, and maintaining this in a very stringent way; translates into latencies of tens of milliseconds to one millisecond and reaching them with very high probability, approximately 99% to 99.999%.
- Logical network resource isolation: public and private 5G systems may share the same physical infrastructure; however, they can't communicate with each other. This might be achieved by network slicing.

- Physical network resource isolation (optional): interworking of physically and resource separated private and public 5G systems.

#### Operation and Management

Operations and management is the means by which operators/owners can run and manage the private 5G systems. It comprises capabilities such as authentication and authorization of management activities, statically or dynamically creating slices, configuration management, scaling, or monitoring (including access to data). In addition, it includes activities for automation and orchestration (e.g., for supporting automated manufacturing processes).

#### Privacy and Security

In private and industrial 5G systems, security aims for the ability to assure decisions (both network operation and data forwarding ones). Data privacy refers to the means to decide where information goes and who can access it. However, different industries require very different security policies, which often determine the deployed network configuration.

The degree of privacy is mainly influenced by the degree of isolation (physical as well as logical) for data, control, and management. The main privacy and security concepts are

- Data privacy through isolation;
- Control and management privacy through isolation;
- Flexibility in choice of security mechanisms;
- Global availability of security mechanisms.

## 12.2 Technology Overview

Private 5G networks offer mobile services to a defined set of users within an organization or for groups of organizations. These p5G systems are typically deployed on the organization's premises, such as a campus or a factory. The p5G systems may have four different and appealing characteristics:

- High quality-of-service requirements, such as ultrahigh reliability, very low latency, or very high throughput.
- High security requirements met by dedicated security credentials that are managed by the organizations that use or benefit from the p5G systems.
- Isolation from other 5G networks as a form of protection against malfunctions in the public system. This isolation, however, might also sup-

port performance, security, privacy, and safety. In addition, some enterprises may view a longer service life than what a commercial carrier may offer, as imperative to their business investment.

- Accountability, in that nonpublic networks make it easier to identify the responsibility for availability, maintenance, and operation.

### 12.2.1 Deployment Scenarios

Initially, p5G systems will focus on factories, IIoT (sometimes just called IoT, and also possibly Industry 4.0), and cyber-physical scenarios. These scenarios, however, comprise very different use cases with a large variety of detailed deployment configurations.

From a high-level viewpoint, private 5G network deployment can be categorized into two classes that differ by ways of isolation:

1. Private 5G networks deployed as isolated and standalone networks;
2. Private 5G networks deployed in conjunction with a public network.

The first p5G class comprises a single configuration, while the second comprises three subconfigurations. These subconfigurations differ in terms of the ways of interacting and infrastructure sharing with the public network. For all these scenarios, it is assumed that all networks provide all services and capabilities required by the p5G at the defined level, and that corresponding SLAs are in place between the p5G operator and one or more public 5G network operators. For example, some aspects of the private network may require features of URLLC and or mMTC that in turn require the use of the 5G SA core, which has been previously discussed in Chapter 2.

There are other factors to be considered when deploying p5G systems. These include, for instance, what radio spectrum is to be used and the spectrum license holder constraints on the frequency use (e.g., power, who owns and operates each network, and what level of trust exists between the p5G operator and the public 5G network operator). In addition, one needs to consider the availability of solution components and the economic feasibility, for example, in terms of total cost of ownership. While these factors are very important, and some of them may be implicitly addressed in the given scenarios, they are beyond the scope of this chapter. Spectrum aspects are discussed, for example, in the white paper from 5G-ACIA, "5G for Connected Industries and Automation" [1].

Figure 12.1 defines the icons depicting the logical elements in the p5G scenarios in the upcoming discussion and diagrams.

| Icon | Description |
| --- | --- |
|  | Device that can communicate via the Radio Network |
| private 5G  | Radio Network only accessible to Non-public network devices |
|  public 5G | Radio Network in Public Network |
| private 5G  public 5G | Radio Network accessible to both Public & Non-Public Network Devices |
|  | User Plane Gateway only accessible in Non-Public Network |
|  | User Plane Gateway in Public Network |
| 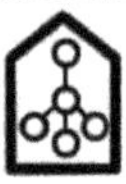 | Control Plane functions in Non-Public Network |
| 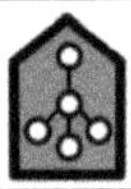 | Control Plane functions in Public Network |
| 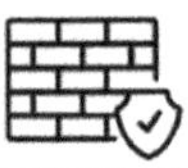 | Firewall |
|  | Subscriber database for Non-Public Network subscribers |
|  | Subscriber database for Private Network subscribers |
|  | Services offered via Public Network, such as Voice and MBB |
|  | Services inside the factory site, such as control and automation systems |

**Figure 12.1** Elements and services in various p5G scenarios.

### Scenario 1: Standalone Nonpublic Network—Isolated Deployment

In the first scenario, the p5G network is deployed in a very isolated way and as a fully independent and standalone network as shown in Figure 12.2. All network functions are located within the logical perimeter of a specific physical premises, for example, a factory. The control of the private 5G system is independent from any control of other 5G systems, although a connection to another public network may exist through a gateway or a firewall node.

The firewall in this scenario forms a clearly defined demarcation point between the private network and the public network. The operational technology (OT) company, which is the company that runs the private 5G system, is fully and solely responsible for operation of the private 5G system. All services offered by the private system are available only within the logical and physical premise boundary.

This type of private 5G network is based fully on the standards and technologies defined by ETSI/3GPP. The system is identified by its own dedicated private 5G network ID. The optional connection to the public network through the firewall can be used to access public 5G network services, such as voice communication to outside 5G systems, while the user is within the coverage area of the private 5G network. In this p5G scenario, devices in the private 5G network may also have dual subscriptions (i.e., they can subscribe directly to the public network). The optional connection to the public systems can be used to bootstrap the devices and to configure and authenticate them for connection to the private 5G network.

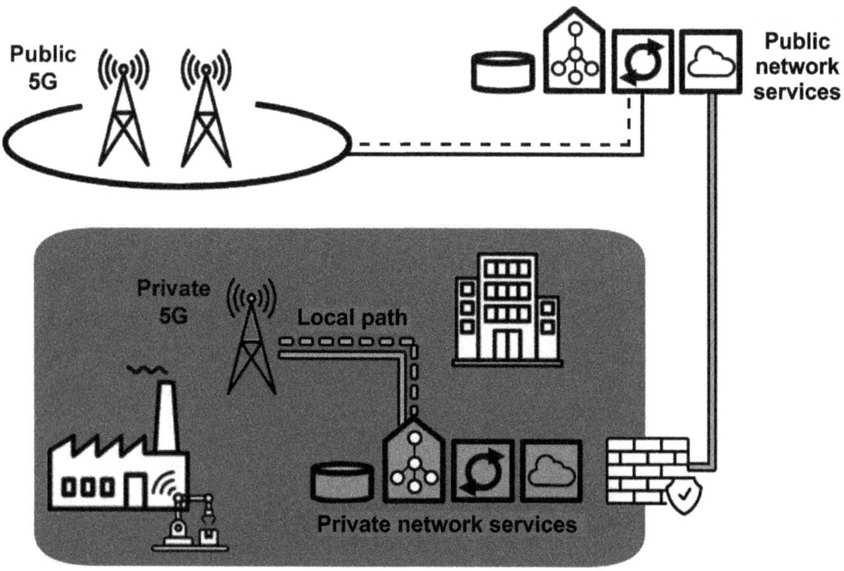


**Figure 12.2** p5G deployment as an isolated network.

Finally, the OT may negotiate roaming agreements with one or more public 5G network operators. This network access may be used for bootstrapping, but roaming agreements may impose technical constraints. The suitability of the scenario depends on the specific use case and their needs for internetworking with other 5G systems. This feature would allow, for example, a worker in the enterprise to have their UE on the private network and fully secured while in the range of the p5G cells, and to seamlessly roam onto a public (or multiple public) networks once outside the range of the p5G cells.

### Scenario 2: Nonpublic Network in Conjunction with Public Networks

This scenario comprises two subconfigurations that are a combination of public and nonpublic 5G networks. The configurations differ by sharing the radio part only, or by sharing both the radio and the control subsystem.

Both subconfigurations assume that certain user services offered within the defined premises are supported entirely by a public 5G network, whereas other services require a dedicated private system. Hence, these scenarios comprise two networks, one public 5G network and one private 5G network. Each network is carrying the appropriate traffic.

#### *Subscenario 2a: Shared Radio Access Network*

The first subscenario is shown in Figure 12.3. Here, the public and the private 5G systems share parts of the radio access network, but the other 5G network functions are kept separate.

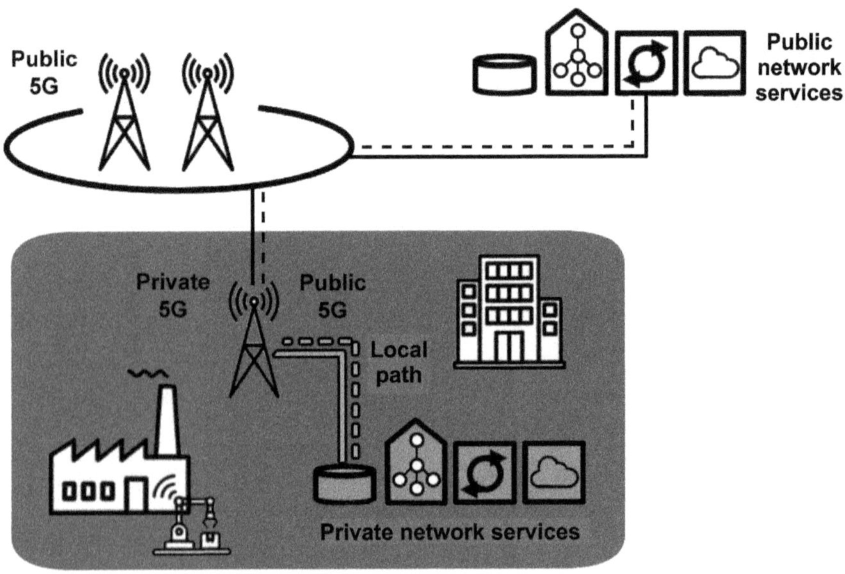


**Figure 12.3** p5G deployment with shared RAN.

In this use case, all data flows of the services of the nonpublic network, either for control data or for user data, are kept within the logical perimeter of the defined physical premise of the p5G system (e.g., within the factory). This enables the OT to control this system entirely.

The part of the 5G traffic that is public is handed over to a public system and is managed there. ETSI/3GPP details this mode in their specification on network sharing [6].

Figure 12.3 shows the deployment with a shared RAN and assumes, for simplicity, only a single base station whereas in practice there may be multiple bases stations in the coverage area of a private 5G system. In this scenario, the base stations are shared for public and nonpublic services for users of the p5G system. The unique user identifier for the private enterprise assures that their traffic remains within the private core, while general public users in the RF footprint of the RAN in the enterprise are treated by the operator's core. This is one of the lesser discussed network slicing models. The RAN may offer different QoS based on the nature of the subscriber, p5G users, or a general user of the network who has roamed into the p5G network space.

As in a public 5G system, the private network uses 5G technology by defined by ETSI and 3GPP but has its own and dedicated private 5G network ID. The scenario requires an agreement with a public 5G operator for sharing the RAN.

Furthermore, the scenario permits an optional connection between the private 5G and public 5G system through a firewall as outlined in Scenario 1 for a p5G deployment as an isolated network. However, for simplicity this connection is not shown in Figure 12.3.

It should be noted here that this scenario has eventually the same needs for a gateway and a firewall to access public service as outlined in Scenario 1.

#### *Subscenario 2b: Shared Radio Access Network and Control Plane*

In the second subscenario, the private and public 5G networks share the RAN for the coverage area of the private system. The public 5G network performs the management and control of the whole system as shown in Figure 12.4. However, the data for private user services is maintained within the logical perimeter and the physical premise of the p5G system. The data flows for public services are transferred to the public 5G system. For example, the UPF functions are separated, allowing the remaining portions of the network to be shared between the enterprise and the operator, with the operator having overall control and responsibility for the nodes.

The major mechanism applied in this scenario is network slicing [7]. Network slicing, previously discussed in Chapter 9, separates the private and public network traffic by using different network slice identifiers [8].

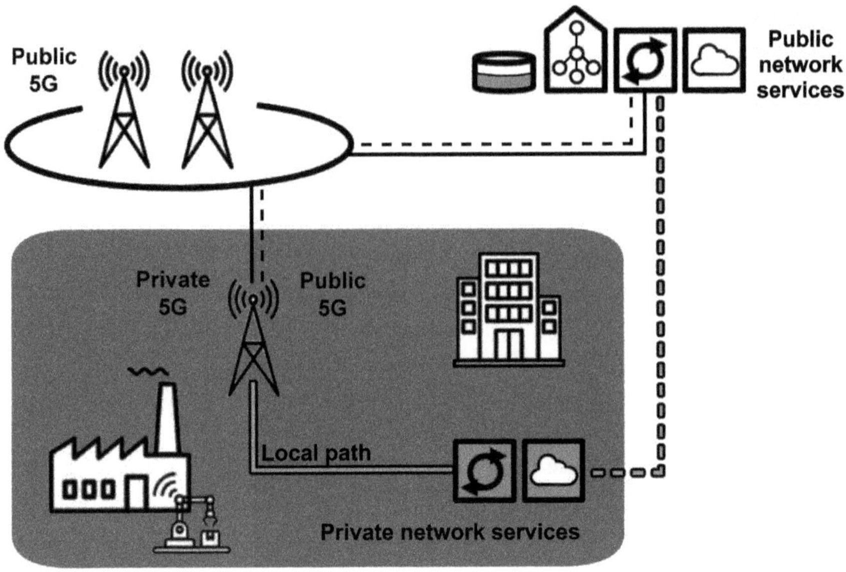


**Figure 12.4** p5G deployment with shared RAN and control plane.

A restricted alternative to implement this scenario, similar to a typical VPN, is to use the 3GPP concept of access point names (APNs) [9]. An APN specifies the target network where the traffic is to be routed to. The APNs are typically software nodes that separate the traffic, but they are typically not as fast and efficient as 5G slicing mechanisms.

Figure 12.4 depicts, for the sake of simplification, a single base station for the RAN on the premises of a factory, instantiating the private 5G system. Additional base stations can be deployed easily and are accessible only to the users of the private 5G network.

The private 5G system is hosted in this scenario by the public 5G network and the devices within the private systems are subscribers to the public network. This simplifies the contractual relationship between private and public network operators and their subscribers. This deployment mode also allows for private 5G devices to connect directly to the public system and its services, including roaming.

This configuration might also comprise an optional connection from the private network services to public network services, as shown in Figure 12.4. Such a connection would permit a coordinated exchange and control of data flows and services. It may also be used to restrict the attachment of private 5G network devices to the public system. If public 5G network services are accessed directly via the public network, the optional connection is not required for accessing network services. The devices can be forced to use the service from the public 5G system.

One concern with this model—and others where the public network hosts the device—is that the ARPU collected by the public operator may not meet that of an individual subscriber to the public network. For example, many enterprises are not willing or able to bear the ARPU cost of each device that connects to the public network. In addition, the public network operator may not have the backend business systems or reporting structure to permit them to segment this market. Clearly, these issues could be addressed if the operators had compelling motivations, but in today's mature market, few are positioned to accommodate this business model at this time.

### Scenario 3: p5G Hosted by the Public Network

The third category of deployment options is when the private 5G network is hosted in full by the public one. This option is depicted in Figure 12.5. Only certain nonpublic 5G services are hosted on premises in this case.

In this option, all 5G user data and control traffic (i.e., both public traffic and private traffic) are handled outside the coverage area of the private system. The private network is treated purely as a logically separate network. Isolation is achieved typically through the virtualization of network functions and the isolation achieved by partitioning the software and computing technology. Here, the virtual functions are executed within operator-managed cloud computing environments. The functions for the public and for the private network are the same regarding their capabilities, but their realization as processes, VMs, or containers in the cloud environment may differ.

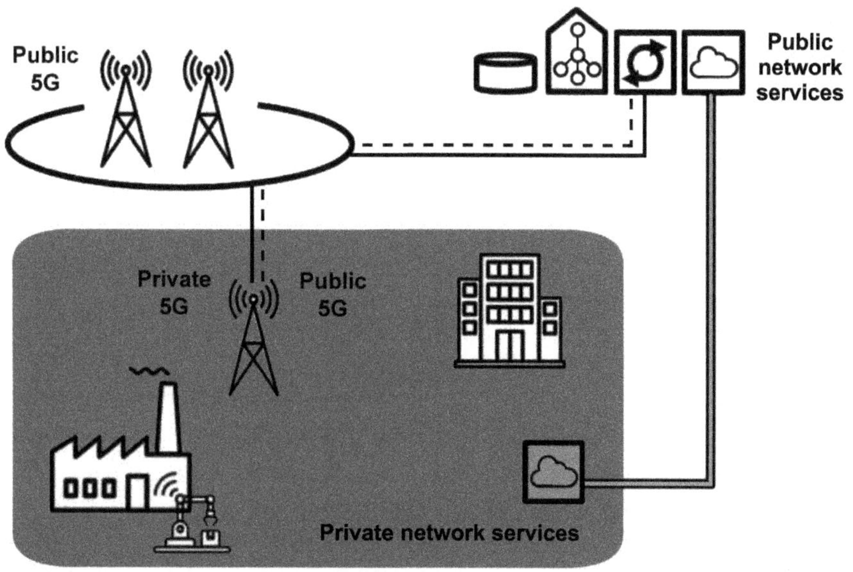


**Figure 12.5** p5G deployed as NPN in a public network.

This hosting scenario for private 5G systems can be implemented by using network slicing or separate APNs. A private 5G subscriber would be defined as a special instance of public 5G subscriber. The scenario further assumes that all data of the private 5G system is routed via the public network. In turn, the access to public 5G services and the roaming for private 5G devices becomes straightforward. Nevertheless, contracts are needed between the public and the private systems to govern the functions and services (e.g., roaming agreements). A dedicated connection as depicted in Scenario 1 between the public and the private systems is not needed in this model.

## 12.3 Multiaccess Edge Computing and Private 5G Systems

MEC is an IT service environment that provides cloud-computing capabilities at the edge of 5G or B5G mobile networks (i.e., within or in close proximity to the RAN). The obvious result is that in an MEC environment the compute resources are close to the mobile subscribers. MEC targets three major objectives: (1) reducing the latency, (2) ensuring efficient network operation and service delivery, and (3) enhancing the user experience for applications (i.e., performance enhancements that are beyond just reducing the latency of packets, such as by enabling service contexts.) MEC is being standardized by an ETSI Industry Specification Group (ISG) of the same name [10].

### 12.3.1 MEC Overview

MEC can be considered to be an evolution of the mobile base station. The concept enables the convergence of telecommunication-focused networking and virtualized IT environments at the edge (i.e., at the boundary between the core of the network and the access point itself). It is facilitated by the enhancement of the capability of the base station. In addition to more capable air interfaces, 5G systems are facilitating programmability concepts that integrate and interconnect software networking and IT virtualization. While the improved latency is a much-needed feature of the 5G spec, MEC offers services, increased scalability, and automation at the edge. These in turn offer the possibility of improved efficiency network operation and service operation.

MEC is based on virtualization platforms that are like the ones of NFV. However, while NFV focuses on network functions and their efficient orchestration and execution, the MEC framework also permits applications to be run at the edge of the network. The virtualization infrastructure that hosts MEC and NFV are by design similar in the foundations and architecture. The similarity enables a reuse of the IT and virtualization infrastructure in both contexts of service (MEC) and network operation (NFV). This reuse should also have implications on the reduction of the OpEx to maintain these systems. For example, the same team may have the necessary skills to maintain the 5G network as the MEC platform. The 5G operators have the clear aim to maximize their infrastructure profitability (i.e., by reducing operational expenditures).

MEC enables the implementation and execution of edge applications as software-only entities that run on top of a capable virtualization infrastructure. The infrastructure is located close or at the base station. Figure 12.6 shows the generic refence architecture for MEC [11]. The architecture is separated into three layers: (1) MEC system level, (2) MEC host level, and (c) networks. In addition, the figures show the reference points between the system entities. The

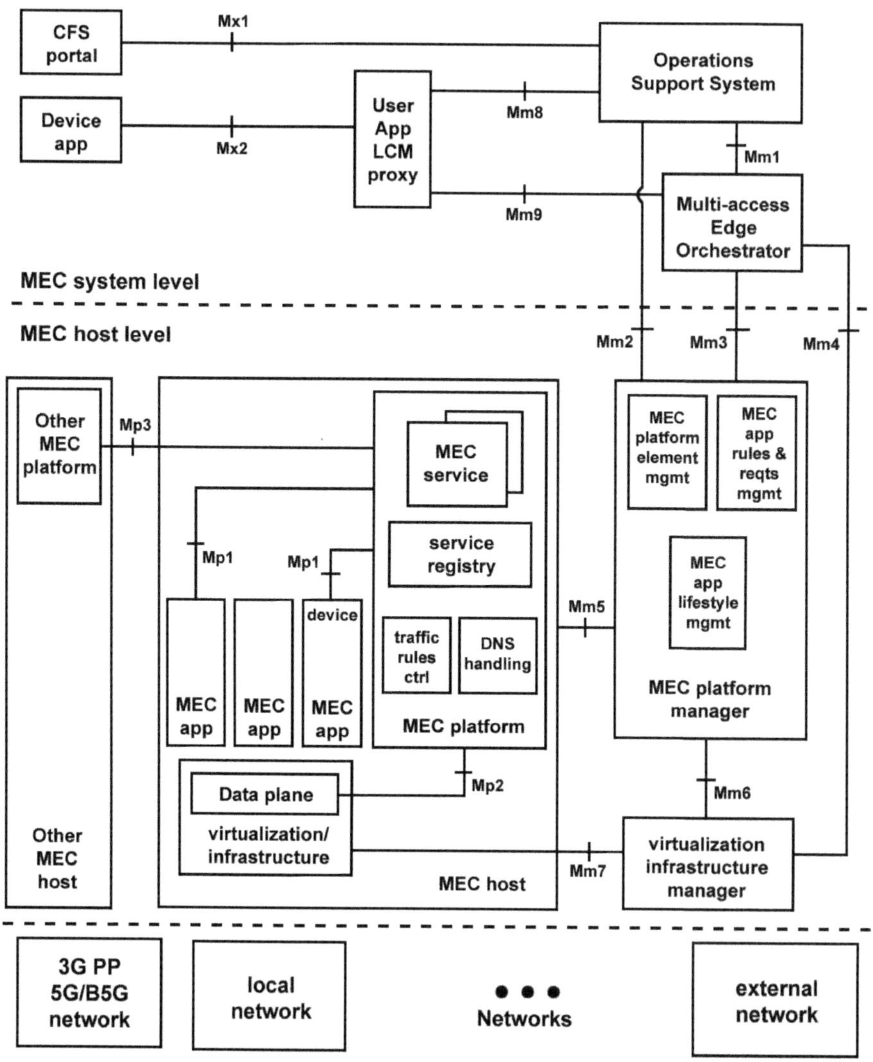


**Figure 12.6** The MEC framework.

elements of the system need to be implemented such that the required functional capabilities at the reference points are achieved.

The categories of reference points are

- Reference points regarding the MEC platform functionality (Mp);
- Management reference points (Mm);
- Reference points connecting to external entities (Mx).

### 12.3.2 MEC Architecture Elements

The most important MEC infrastructure element is the MEC host, as shown in Figure 12.6. This is the entity that manages the infrastructure that virtualizes the physical compute, storage, and network resources of the host for the purpose of running MEC applications.

The MEC platform is a collection of essential functionalities required to run MEC applications on a particular virtualization infrastructure. This enables the host to provide and consume MEC services and support the interface points between modules. MEC applications are instantiated on the virtualization infrastructure of the host based on configurations or service requests validated by MEC management.

MEC management comprises the MEC system-level management and the MEC host-level management. The MEC system-level management includes the multiaccess edge orchestrator as its core component. This element maintains an overview of the complete MEC system. The MEC host-level management comprises the MEC platform manager and the virtualization infrastructure manager and handles the management of the MEC-specific functionality of a particular MEC host and the applications running on it.

All in all, the MEC enables the merging of network and application service provisioning at a location close to the user or the service consumer. It provides a single architecture for IT virtualization and cloud-computing concepts.

### 12.3.3 Future MEC Solutions for Private 5G Systems

The overlap in technologies of private 5G networks and MEC is rather appealing for future applications and the future network architectures supporting them. In addition, ETSI and 3GPP has worked to synchronize the standardization of both technology families. This coordination is visible when considering the documents for the recommended general MEC architecture (e.g., [11]). It explains in Section 6.2 how the MEC architecture and its interfaces integrate and support the NFV interfaces.

However, MEC is still a very young concept in the context of networking and when combined with the use of private networks, various deviations to the intended implementation are not unheard of. The interworking of private networking with private clouds with public clouds is available using commercial solutions, such as AWS hybrid cloud services [12, 13]. Even highly distributed and almost P2P-based private and secure clouds for specific software-development concepts (e.g., for AI systems engineering) are currently gaining popularity in research [14, 15] and are being adopted by smaller companies [16]. An eventual economical challenge might arise in the competition of technologies developed by the larger hyperscalers such as Google, Facebook, and Amazon, and by the technologies favored by the network operators. One concern that has been raised is the difficulty at times to monitor and control costs when placing workloads into the hyperscaler environments. We'll not be providing further analysis at this time on this observation.

However, major technical challenges in combining MEC with 5G systems, and particularly with private ones, are probably buried in the methods and techniques for the separation, reorganization, and orchestration of network functions that control user mobility across networks (e.g., between private 5G networks of the same enterprise but in different coverage areas and interconnecting dynamically private clouds at various locations). Other technical challenges are the provisioning of network services for URLLC, massive machine-type communication, or the selection of the access technology.

An outstanding example on how to address some of these issues is the Taiwanese 5G project 5G Intelligent A+ that is reflected in Chunghwa Telecom's product of the same name. A project that was featured in 2021 and 2022 in the *IEEE Communication Standards Magazine* [17, 18]. The project suggested initially an efficient tunneling concept for interconnecting base stations with on-site, off-site, and other public or core network operator clouds. The generic architecture and tunnel management of the solution is depicted in Figure 12.7.

The solution shown in Figure 12.7 covers the premises of a factory or an enterprise with installed base station facilities. It interconnects these facilities with various clouds (private, public, on-site, or off-site) by sophisticated and adaptive tunnels or leased lines. The architecture's core element is denoted "mobile edge enabler" that coordinates, orchestrates, and optimizes the use of the tunnels.

The architecture can be applied to any kind of radio access network, including Wi-Fi, 4G LTE, and 5G. For 4G, the eNB is connected through the EPC with CP. For 5G NSA Option 3x networks [19] the base stations consist of eNBs and 5G gNBs. Last, for 5G SA networks, the 5G gNB is exclusively used by the UE.

Finally, private and nomadic 5G networks have been extensively researched. They are increasingly considered for disaster relief scenarios. The German Advanced Low Altitude Data Information System (ALADIN) project, for

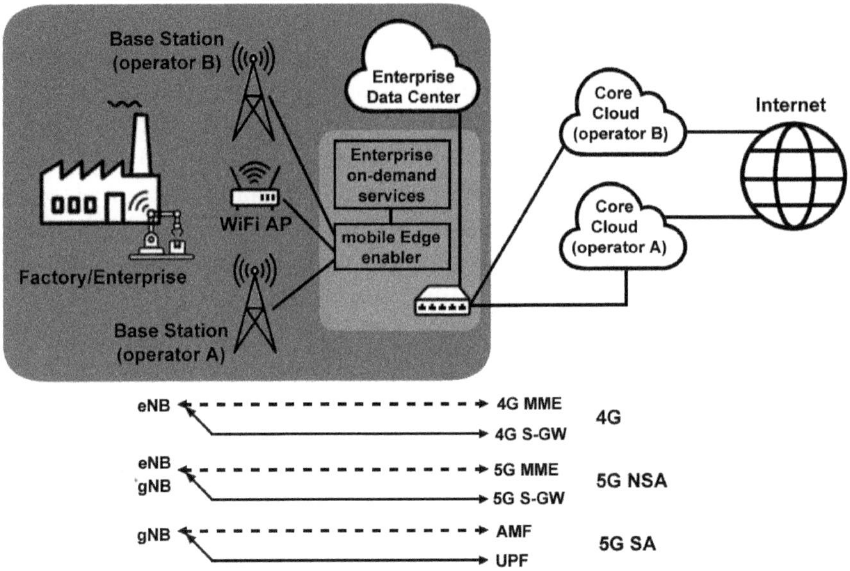


**Figure 12.7** Architecture of the 5G Intelligent A + private 5G interconnection.

example, suggests 5G private networks for firefighting scenarios and puts the base station facilities into an airborne drone [3]. Similarly, ZTE installs private 5G BS facilities on wheels [20]. The major challenges here are to miniaturize the equipment and to manage the mobility of the 5G infrastructure facilities. Mobility imposes additional timing loads on the control loops in the 5G control plane.

## 12.4 Business Issues with Private 5G and MEC Systems

The business issues for 5G systems are expected to be different from previous generations of network. The difference is in part due to the increased openness of 5G systems. While the business issues behind future applications eventually require an application-domain-specific discussion, we discuss the business issues relating to the intersection of hardware and software-based infrastructures for public and private 5G systems and the impact on applications and services. This discussion begins with the economic factors that enable the benefits for private 5G for application and continue to the discussion of the SIM and eSIM cards. The SIM card discussion is significant as the UE requires some form of binding to a specific network, and the SIM (or eSIM or iSIM) provides that binding. Later in this chapter we take this up in more detail.

### 12.4.1 Enabling Private 5G Benefits for Applications

First, in many cases while there is an interest at the technical level, there may be a lack of a business model for private 5G. For example, if there is an existing fiber connectivity option, then the p5G might only be considered as a backup option. As one architect (LJH) is known to say, "if you have fiber, use the fiber." Clearly, in many cases there is no fiber available; sometimes this is due to the device being mobile, and in other cases it may be due to physical constraints such as the location of a fixed device making a fiber drop impracticable.

One current trend is also to consider the use of p5G to close the digital divide. These use cases are often found in one of two environments. The first is in very dense urban residential communities where operators have not deployed, or not deployed at sufficient scale, RAN resources. In many cases unfortunately we find this to be a hard-nosed economic decision by the mobile operators. This reality was driven home by the Covid-19 pandemic where many dense urban schools went virtual, and many students did not have access to the necessary broadband connectivity to effectively participate in the e-learning model that was implemented. This was even noted by the local school board in the home city of one of the world's largest mobile operators [21].

We also see interest in p5G for the home health care needs of people in these underserved communities, either in urban or rural settings. Private 5G, while attractive technically, raises business and operational concerns in these cases. On the operational side there is the need for ongoing maintenance of the network, and many institutions may be unprepared for the demands and cost of maintaining a radio network and the associated core elements. Second, just popping up a radio antenna may require permitting and regulatory compliance that is not easily obtained in every case. For example, while a local school district placing an antenna at an elementary school may at first be appealing (provided the school itself has the necessary broadband connectivity to provide the transport back into a core network to access the internet), the network buildout and RF engineering may represent a significant cost that is not easily recovered. There is a similar consideration in the health care use case. The demand is clearly there and the technology exists, but the reality is the health industry is very likely unable to themselves acquire the real estate needed for the location of the necessary radios.

A second domain of interest is the resource-dense, people-sparse spaces that have been mentioned earlier. These environments may lack fiber to the end point device, or have a mobility requirement of the devices or some other blocker preventing hard connections to the devices. In addition, the volume of these devices prevents a hard fixed connection, such as a cable or fiber, from being used.

There is a third popular domain, and the use case is found in dense population venues such as sporting or other entertainment venues, where the density of the mobile users would exceed the typical capacity of a macrocellular base station, but only when an event is in progress. In these last two domains the environment is suited to the installation of the radio equipment needed for the anticipated workload. In the previous two examples, the actual deployment of the radios may be the blocker that prevents any further consideration; for example, where there is no RF license there can be no p5G.

Often, the business partner showing interest in p5G is not really seeking a 3GPP-compliant radio solution, but rather they are seeking a business outcome and believe for some reason that part of the solution to achieve that outcome requires p5G. While this may be true in many cases, there are also cases, such as an electric meter, where some have proposed using p5G in this use case but there currently are better technologies available and that have been extensively deployed. The point here is that p5G may only be useful for larger businesses who can justify the deployment and ongoing operations of a p5G infrastructure.

We should also address the reasons someone may choose to deploy p5G rather than a Wi-Fi solution. In doing so, we freely recognize that there are cases where the combination of a 3GPP cellular wireless technology along with Wi-Fi may be the preferred approach. These hybrid networks are ideal in brown-field edge or enterprise deployments. However, our focus is on p5G, but does not exclude an overlay where Wi-Fi may be included as part of a solution.

The 3GPP network solution with a p5G can have several economic and operational benefits over a Wi-Fi-only solution. First, some studies have shown that the TCO of a 3GPP solution in large manufacturing enterprise applications can be lower than that of a Wi-Fi solution. The number of access points (RANs) can be significantly reduced, p5G simplifies the spectrum management if and when new access points are introduced, and the increased security offered by the cellular network can all be advantages. p5G may also allow significantly more devices to be connected to any single access point. In addition, in very large enterprise environments, the possibility of using either dedicated or dual SIM devices can significantly improve the experience of workers on the floor who may have a large roaming work area, such as maintenance workers.

Some studies have broken down the current private wireless into four provider quadrants. These are (1) carrier provided, (2) system integrator provided, (3) NEP/TEM provided, and (4) enterprise led. In one study by Omdia (a private report) [22], in November 2022 the carriers were lagging behind the leading systems integrators, network equipment providers, and enterprise in total volume and revenue, with over 70% of the market and the operators with just over 20% of the market. Clearly, if the operators are going to monetize 5G by expanding into the enterprise space this trend will need to change.

### 12.4.2 SIM, eSIM, iSIM

The onboarding of devices onto a p5G network still requires the device to establish a secure connection and complete registration with the core of the network, even if that core resides at the edge. This requires the use of a SIM, and while many are likely familiar with the SIM card found in the typical handset today, using a physical device that contains the necessary network and identification information to establish a binding with a specific network is impractical for the enterprise user and the IoT devices likely to reside on the network. To address this, the embedded SIM (eSIM) was introduced in 2015 and has seen some limited use in handsets until recently with the introduction of higher-end iPhone and Android devices. On today's IoT devices the eSIM is still a SIM card, it's just not possible for the end user to remove the SIM card via an accessible slot on the device. The use of the eSIM in nonhandset devices still requires some form of configuration and binding on the device. Devices that do not have a native input capability such as a keyboard and possibly camera to scan a QR code present a particular challenge that is currently a concern for operators. Some device manufactures have different solutions to configure or otherwise provision the eSIM. This raises the difficulty level of scaling operations for the providers. Recent introduction of the integrated SIM (iSIM) reduces the space in the device and reduces power consumption over the SIM and eSIM options. The unification of the onboarding of the devices, whether through the use of an eSIM or iSIM remains an area of concern not only for operators but also for enterprises considering large scale (i.e., tens of thousands or more) of connected devices from various manufacturers.

### 12.4.3 MEC and Hyperscalers at the Edge

It was originally envisioned that the MEC resources would be installed only at the base station. Today we have found that many enterprise applications have a need for these resources to be on-premises or at the enterprise location themselves. This can be for a number of reasons, many of which are, if not identical, then parallel to the fundamental response the p5G is being used, other than closing the digital device of lacking active spectrum. While the operators would ideally see the MEC being deployed in their network this is not always the clear choice.

MEC is a framework, and there are several MEC-like offerings both in network operators today and from the NEP and SI communities. Some examples are the Nokia Digital Automation Cloud (NDAC), Ericsson Private networks, Mavenir's Access Edge Solutions, WWT's Private LTE solution, HCL's Private LTE/5G, and Tech Mahindra's Private Networking for Enterprises,

Intel's SmartEdge™, to name only a few of the available commercial offerings. Here we have omitted a list of references, but a quick search should provide an expansive list of these and other commercial resources in the private networking space.

Microsoft [23], as mentioned in Chapter 1, has acquired two elements of the mobile core with their purchase of Affirmed [24] and Metaswitch in 2020 [25]. In addition, Microsoft has been a key partner to AT&T [26] since mid-2021 to operate their 5G core in Azure.

AWS [27] is currently likely to appear as the purest instance of a private 5G offering from the three hypserscalers.

Google [28] in some markets uses their Distributed Cloud and a network slice (where available) to allow enterprises to deploy applications closer to the mobile edge. This provides the enterprise with what Google terms a dedicated network channel into the operator's network, providing optimized bandwidth, improved latency, and higher reliability, as well as improved security. This level of isolation and deployment of the enterprise application closer to the edge may be a stretch for the purist in classifying this as a private 5G network, but it clearly maps well to the MEC model, and we would not object to this being included in a discussion of private networking.

## 12.5 Summary

This chapter discussed and detailed the concepts of the private 5G network and MEC. Both concepts make 5G systems more open to new usage scenarios, user communities (e.g., industries), and service and application provisioning (e.g., services at the edge). These additions to 5G demonstrate the flexibility of the 5G concept and the increased convergence of transmission, technology, networking, computing, and applications. In addition, we outlined specific business issues to provide a better understanding of how to design and operate p5G and MEC network-compute concepts.

To leverage economies of scale, clouds, and cloud technologies for p5G and MEC need to avoid becoming silos or telco-specific clouds. The use of standard cloud systems is likely not technically detrimental to 5G and p5G network operation; in fact, their use could provide the advantages of flexibility and innovation [29].

While p5G and MEC focus on typical networking and computing improvements, such as mobility support for haptic applications or URLLC services, the energy efficiency of combining 5G, p5G, and MEC is limited. A holistic approach is needed that considers the energy consumption of both the computing and the networking parts of public and private 5G systems [30–32].

## References


[1] *5G Non-Public Networks for Industrial Scenarios,* white paper, https://5g-acia.org.

[2] ETSI/3GPP, 5G; Service Requirements for Cyber-Physical Control Applications in Vertical Domains, Technical Specification (TS) 22.104 version 17.7.0 (Release 17).

[3] ALADIN, Advanced Low Altitude Data Information System, https://aladin-5g.de.

[4] Bai, C., P. Dallasega, G. Orzes, and J. Sarkis, "Industry 4.0 Technologies Assessment: A Sustainability Perspective," *International Journal of Production Economics,* Vol. 229, September 2020.

[5] ETSI, "Multi-access Edge Computing," https://www.etsi.org/technologies/multi-access-edge-computing.

[6] ETSI, Universal Mobile Telecommunications System (TMTS); LTE; Network Sharing; Architecture and Functional Description (3GPP TS 23.251 Version 13.1.0 Release 13), 2016, https://www.etsi.org/deliver/etsi_ts/123200_123299/123251/13.01.00_60/ts_123251v130100p.pdf.

[7] ETSI/3GPP, Group Services and System Aspects; Management and orchestration; Concepts, Use Cases and Requirements, Technical Specification (TS) 28.530 Version 17.3.0 (Release 17).

[8] ETSI, 5G; 5G System; Network Slice Selection Services; Stage 3 (3GPP TS 29.531 Version 16.3.0 Release 16), https://www.etsi.org/deliver/etsi_ts/129500_129599/129531/16.03.00_60/ts_129531v160300p.pdf.

[9] ETSI/3GPP, Group Core Network and Terminals; Numbering, Addressing and Identification," Technical Specification (TS) 23.003 Version 17.7.0 (Release 17).

[10] ETSI, Multi-Access Edge Computing (MEC), https://www.etsi.org/technologies/multi-access-edge-computing.

[11] ETSI, Multi-access Edge Computing (MEC); Framework and Reference Architecture, Group Specification (GS) MEC 003 Version 3.1.1 (Release 17), https://www.etsi.org/deliver/etsi_gs/MEC/001_099/003/03.01.01_60/gs_MEC003v030101p.pdf.

[12] Amazon, Hybrid Cloud with AWS, https://aws.amazon.com/hybrid/.

[13] Microsoft, Azure; Invent with Purpose. Learn, Connect and Explore, https://azure.microsoft.com/.

[14] Ahmadi Mehri, V., *Towards Secure Collaborative AI Service Chains,* Licentiate Dissertation, Blekinge Tekniska Högskola, Karkslrona, September 2019.

[15] Tkachuk, R.- V., *Towards Decentralized Orchestration of Next-generation Cloud Infrastructures,* Licentiate Dissertation, Blekinge Tekniska Högskola, Karkslrona, June 2021.

[16] Llewellynn, T., M. M. Fernández-Carrobles, O. Deniz, et al., "BONSEYES: Platform for Open Development of Systems of Artificial Intelligence," in *Proceedings of the Computing Frontiers Conference (CF'17),* ACM, 2017.

[17] Kao, L.- C., and W. Liao, "5G intelligent A+: A pioneer Multi-Access Edge Computing Solution for 5G Private Networks," *IEEE Communications Standards Magazine,* May 2021.

[18] Kao, L.- C. and W. Liao, "Multi-Access Intelligent A+: Ensuring Service Continuity and Consistency in Private 5G Heterogeneous Networks,," *IEEE Communications Standards Magazine,* June 2022.

[19] 3GPP, *Group Services and System Aspects; Release 15 Description; Summary of Rel-15 Work Items,* Technical Report (TR) 21.9015 Version 15.0.0 (Release 17), September 2019https://portal.3gpp.org/desktopmodules/Specifications/SpecificationDetails.aspx?specificationId=3389.

[20] LightReading, "All-in-One Nomadic 5G New Telecommunication Solutions for Rapid Response in Disaster Zones and Beyond," October 17, 2022, https://www.lightreading.com/all-in-one-nomadic-5g-new-telecommunication-solutions-for-rapid-response-in-disaster-zones-and-beyond/a/d-id/781136.

[21] Vaughn, J., "Dallas Tries to tackle Digital Divide," *Dallas Observer,* August 9, 2021, https://www.dallasobserver.com/news/dallas-tries-to-tackle-digital-divid-12160722.

[22] Omedia, https://omdia.tech.informa.com/insights/5g-marketing-insights-lp.

[23] Microsoft, "What is Azure Private 5G Core Preview," October 27, 2022, https://docs.microsoft.com/en-us/azure/private-5g-core/private-5g-core-overview.

[24] Khalidi, Y., "Microsoft Announces Agreement to Acquire Affirmed Networks to Deliver New Opportunities for a Global 5G Ecosystem," *Official Microsoft Blog,* March 26, 2020, https://blogs.microsoft.com/blog/2020/03/26/.

[25] Khalidi, Y., "Microsoft Announces Definitive Agreement to Acquire Metaswitch Networks, Expanding Approach to Empower Operators and Partner with Network Equipment Providers to Deliver on Promise of 5G," *Official Microsoft Blog,* May 14, 2020, https://blogs.microsoft.com/blog/2020/05/14/.

[26] Microsoft News Center, "AT&T to Run Its Mobility Network on Microsoft's Azure for Operators Cloud, Delivering Cost-Efficient 5G Services at Scale," https://news.microsoft.com/2021/06/30/att-to-run-its-mobility-network-on-microsofts-azure-for-operators-cloud-delivering-cost-efficient-5g-services-at-scale.

[27] Amazon, "AWS Private 5G," https://aws.amazon.com/private5g.

[28] Jain, A., and Sat K, "Google Delivers 5G Network Slicing Capabilities for Enterprises," https://cloud.google.com/blog/topics/telecommunications/5g-network-slicing-with-google-android-enterprise-and-cloud.

[29] Sbeglia Nin, C., "'That Isn't the Point': Why AWS Doesn't Use the Term Telco Cloud," *RCR Wireless News,* October 12, 2022, https://www.rcrwireless.com/20221012/telco-cloud/that-isnt-the-point-why-aws-doesnt-use-the-term-telco-cloud?

[30] Wen, M., Q. Li, K. J. Kim, et al., "Private 5G Networks: Concepts, Architectures, and Research Landscape," *IEEE Journal of Selected Topics in Signal Processing,* Vol. 16, No. 1, 2022.

[31] Simsek, M., A. Aijaz, M. Dohler, J. Sachs, and G. Fettweis, "5G-Enabled Tactile Internet," *IEEE Journal on Selected Areas in Communications,* Vol. 34, No. 3, 2016.

[32] Chukhno, O., O. Galinina, S. Andreev, A. Molinaro, and A. Iera, "Interplay of User Behavior, Communication, and Computing in Immersive Reality 6G Applications," *IEEE Communications Magazine,* Vol. 6, No. 12, 2022, pp. 28–34.

# 13

# Open-Source Software Development and Experimental Activities

## 13.1 Introduction

The research community has responded to the many challenges of designing and developing high-performance 5G technologies in two ways. On the one hand, it has focused on the development of open-source 3GPP software implementations and deployment frameworks that can be effectively used to quickly build 5G PoC designs. On the other hand, it has pursued the construction of publicly available open experimental labs, which offer researchers time-shared access to various 3GPP technology services. These ongoing efforts are reviewed in this chapter.

## 13.2 5G Open-Source Software Packages

This section provides an overview of open-source 3GPP software implementations and some deployment frameworks that can be used to build PoC designs. In comparison to costly proprietary and closed-source 3GPP-compliant software, open-source solutions (when combined with readily available laboratory RF hardware) allow experimenters to build low-cost small-scale PoC designs that even allow the use of commercial cellular terminals (user equipment) as opposed to emulators or nonstandards-compliant equipment. In addition, the open-source approach facilitates the integration of commercial (i.e., vendor) network elements.

### 13.2.1 Open-Source 5G Core Network Elements

The core network in a 3GPP radio system requires a collection of protocols that are needed to interconnect radio-access networks and provide services such as telephony and internet access to mobile users. These protocols and related interfaces are defined to implement both the control plane and user plane. This section reviews several popular open-source software packages that are often used to implement 5G core network PoC designs.

#### 13.2.1.1 OpenAir-CN-5G

The goal of the OAI-5G CN project group is to provide a 3GPP-compliant 5G SA CN implementation with a rich set of features. Currently, OAI 5G CN supports basic procedures for connection, registration, and session management. It also supports features such as network slicing (partial support), N2 handover, HTTP/2, FQDN, and paging support. Each feature of the OAI 5G CN components is continuously tested with professional testers, commercial gNBs (with COTS user equipment), and open-source RAN simulators. It supports minimalist 5GC deployment, basic 5GC deployment, and slice-based deployment, the details of which are available in [1]. The OAI-5G-CN can run on bare metal or in virtual machines. In addition, the network functions can be deployed as Docker containers using Docker-compose files. To support cloud-native deployment, OAI 5G CN can be installed using a Helm chart on openshift or Kubernetes clusters. The code is distributed via GitHub (https://gitlab.eurecom.fr/oai/cn5g/oai-cn5g-fed/-/tree/master) under an Apache V2.0 license and is maintained by the OpenAirInterface Software Alliance (OSA), which provides a CI/CD testing framework for community contributions to the code base.

#### 13.2.1.2 Open5GS

Open5GS offers an open-source implementation of a Release 16 4G/5G NSA-compliant EPC, and a 5G SA core (i.e., the core network of LTE/NR network). One can build their private network using Open5GS if gNB/eNB and USIM are available. The 4G/5G NSA network components and 5G SA network functions are written in C and distributed under the GNU Affero General Public License (AGPLv3). The web user interface (UI) is available for testing purposes and is implemented in Node JS and React. Commercial licenses are also available from NextEPC, Inc. It is compatible with various Linux distributions, such as Debian, Ubuntu, Fedora, and CentOS, as well as FreeBSD and macOS. The code is distributed via GitHub link https://github.com/open5gs/open5gs and user guide documentation is available at [2].

#### 13.2.1.3 free5GC

free5GC is an open-source project for 5G mobile core networks. The goal of this project is to implement the 5GC defined in 3GPP Release 15 (R15) and beyond. fee5GC is written in the Go programming language and is compatible with equipment running the Ubuntu Linux operating system. The code is distributed via GitHub (https://github.com/free5gc/free5gc) under Apache V2.0 license and is maintained by NCTU.

### 13.2.2 Open-Source Evolved Packet Core

The EPC is a standard framework that provides converged voice and data by replacing circuit switching with packet-switching technology. EPC powers 4G LTE networks and 5G non-standalone networks. This section lists some open-source software packages that may be used to run an EPC.

#### 13.2.2.1 OMEC

OMEC claims to be the first full-featured, scalable, high-performance open-source EPC. OMEC is built using an NFV architecture with 3GPP Release 13 compatibility. It is developed primarily by an operator-driven consortium and is an industry-led open-source community. The primary contributors are Sprint (now T-Mobile), AT&T, Intel, and Deutsche Telekom. It supports many subscribers with a high-performance DPDK-based data plane. It is optimized to handle lightweight cost-effective deployments and IoT applications. It is designed to be used as a standalone EPC and is also an upstream project for the COMAC platform [3] that integrates both mobile and fixed subscriber management functions.

#### 13.2.2.2 Facebook Magma

Magma is an open-source project maintained as a product by Facebook (Meta) Connectivity. It is a small-scale EPC augmented by network management, orchestration, and performance metric measurement functions. In addition, it provides a federation gateway to interface with 3GPP standardized interfaces in operators' existing core networks. It is designed to be deployed in a cloud-native environment making use of the VirtualBox hypervisor and Vagrant. This open-source software platform's main target is to enable operators to easily deploy mobile networks in rural or remote areas of the globe that do not have 4G coverage. Magma is distributed under the BSD License, as found in the LICENSE file in the Git location https://github.com/magma/magma.

#### 13.2.2.3 srsEPC

srsEPC is a lightweight implementation of a complete LTE core network (EPC). The srsEPC application runs as a single binary but provides the key

EPC components of home subscriber service (HSS), mobility management entity (MME), service gateway (S-GW), and packet data network gateway (P-GW). The EPC implementation included in srsEPC is written in C++. It is compatible with the Ubuntu and Fedora Linux operating systems and can be installed using the srslte repository. The code is available at https://github.com/srsran/srsRAN/tree/master/srsepc and user manual documentation can be found in [4].

### 13.2.3 Open-Source Radio Access Network Elements

The gNB in the 5G RAN is split into a CU and DU. These two modules can be disaggregated and distributed across a wide-area network or within a data center. This split allows the DU to perform lower layer functions (RLC layer and below) in smaller geographic regions (small cell or microcell radius of up to a few kilometers) and the CU to perform aggregated higher-layer services over large regions (approximately up to a 200-km radius). There are many open-source RAN elements available at this time and some are briefly reviewed next.

#### 13.2.3.1 OpenAirInterface 5G RAN

OpenAirInterface 5g is an open-source development community producing a full software implementation of the 3GPP radio-access network components. The scope of the OAI 5G RAN project is to develop and deliver a 5G software stack under the OAI Public License V1.1. The following options are available in the OAI 5G stack:

- NSA gNB;
- SA gNB;
- 5G NSA and SA UE.

In the SA mode, the available CU/DU split deployment is validated with the OAI RF simulator. The RF simulator simulates the L1 layer, which replaces the physical radio board with software (TCP/IP) communication that enables all functional tests without requiring an RF board. It also supports the commercial RF board USRPs and LimeSDR. The performance of CU/DU split using real RF and COTS UE are currently being validated. The CU-C (control plane)/CU-D (data plane) split over the E1 interface is currently integrated with the support of multiple CU/DUs. Additional features support MIMO, T1-Offload, and L2 simulation for research purposes. Although not mentioned explicitly in the development plan, RIC is an important component of OpenAirInterface and its developments are part of the OAI MOSAIC5G Project Group. The OAI 5G RAN roadmap is available at [5] and Git code repository links are available at [6]. The RAN is licensed under the OAI Public License V1.1. The OAI Public License V1.1 is a modified version of Apache V2.0 License, with the modified patent clause that allows contributing parties to make patent licenses available to third parties under fair, reasonable, and nondiscriminatory (FRAND) terms for commercial exploitation. OAI code is free for noncommercial/academic research purposes.

#### 13.2.3.2 srsRAN

srsRAN is a free and open-source 4G and 5G software radio suite [7]. Featuring both UE and eNodeB/gNodeB applications, srsRAN can be used with third-party core network solutions to build complete end-to-end mobile wireless networks. srsRAN is released under the AGPLv3 license and source code is available at https://github.com/srsran/srsRAN. It offers LTE Release 10 functionality aligned with features up to Release 15 and interoperates with software-defined radio devices such as USRPs and LimeSDR. The srsRAN feature list is available at [8].

#### 13.2.3.3 O-RAN Amber

O-RAN Amber [9] is a partially open-source software suite for RAN prototyping and proof-of-concept (PoC) demonstration of the O-RAN interface specifications. It was released in December 2019 and currently contains a 4G implementation and some elements of the 5G SA protocol stack. The software includes both RAN implementations based on the O-RAN license (FRAND license) and closed binaries implementing physical layer procedures. It also includes initial implementations of the so-called RIC, which are released under an Apache V2.0 license.

### 13.2.4 Open SDR Devices

The RF Layer 1 functionality is achieved with the help of radio boards. The canonical example of software-defined radio device is the Universal Software Radio Peripheral (USRP) family originally commercialized by Ettus Research [10], which is now a subsidiary of National Instruments [11]. The Ettus devices are generally of high quality in terms of RF performance and software stability. Ettus maintains an open-source driver and development environment (USRP Hardware Driver (UHD)) for interfacing with the USRP family of devices. UHD is licensed with a GPL V3.0 for nonprofit research or commercial implementations making use of compatible licensing technology. USRP devices supported by the open-source RAN packages include USRP B200, B210, X300, N300, and N310. These SDR devices offer a range of interfaces, capabilities, and prices.

Nuand [12] produces another family of similar low-cost prototype devices known as the BladeRF. In terms of its RF and acquisition technology, the

BladeRF 2.0 is essentially the same as the B210 device from Ettus in a different package and with a different software API to interface with the host computer.

Finally, the LimeSDR family of devices makes use of RF and acquisition technology based on silicon from Lime Microsystems.

### 13.2.5 Open-Source Control and Orchestration

Resource control and orchestration solutions can be investigated using an ecosystem of platforms and use cases for 4G/5G system research that typically comprises SDN, NFV, cloud-native, and MEC technology enablers.

#### 13.2.5.1 OAI Mosaic 5G

The OAI Mosaic 5G (M5G) Project Group aims to transform radio access (RAN) and core networks (CN) into agile and open network-service delivery platforms. The focus of the Mosaic 5G program is on providing software implementations of flexible the RAN intelligent controller called FlexRIC, a RAN agent that enables interfacing with the radio stack and a real-time (RT) controller. The functionalities in the FlexRIC RT controller component are similar to those in the O-RAN Alliance's near-RT RIC. Some other features include Trirematics [13], an intelligent RAN and CN operator, and flexCN, a flexible core controller. The detailed roadmap of OAI Mosaic 5G is available in [14].

#### 13.2.5.2 Akraino

Launched in 2018, Akraino [15] Edge Stack aims to create an open-source software stack that supports high-availability cloud services optimized for edge computing systems and applications. The Akraino Edge Stack is designed to improve the state of edge cloud infrastructure for enterprise edge, over-the-top (OTT) edge, and carrier edge networks. It offers users new levels of flexibility to scale edge cloud services quickly, to maximize the applications and functions supported at the edge, and to help ensure the reliability of systems that must be always up and running (near-RT RIC).

## 13.3 5G Experimental Networks for US-EU Collaboration

This section provides an overview of currently publicly available open experimental sites which provide 3GPP technology services to researchers at large. More specifically, this section offers a summary and roadmap of some of the most popular experimental sites in the United States (namely, POWDER, Colosseum, COSMOS, and AERPAW) and Europe (namely, 5G-EVE, 5G-VINNI, 5GENESIS, the R2LAB facility in France, and NITOS in Greece), respectively.

### 13.3.1 POWDER

Platform for Open Wireless Data-Driven Experiment Research (POWDER) is a flexible infrastructure that supports software-programmable experimentation on 5G—massive MIMO, O-RAN, spectrum sharing, RF monitoring, and Citizens Broadband Radio Service (CBRS)—that can be achieved on software-defined radios. POWDER is run by the University of Utah in partnership with Salt Lake City and the Utah Education and Telehealth Network. The goal of the POWDER team is to facilitate and expedite the transition of a concept from the research lab into the real world. To realize the vision of city-scale experimentation, the POWDER team has placed dozens of fixed base stations over an area of 14 square kilometers, with 60 mobile devices traveling on couriers and 70 sensor nodes, either fixed or mobile, deployed throughout the area. This contiguous space covers three distinct environments: a high-rise urban downtown, a moderate-density residential area, and a limited-scale radio network on the University of Utah campus. The outdoor area where the RF equipment is deployed is the University of Utah campus. All base stations are located on the campus rooftops, and most are colocated with commercial cell tower equipment to re-create an environment similar to a production-level network. All base stations are connected to the data center through dark fiber [16]. For experiments involving user mobility in the field, the campus has several bus routes that use mobile carriers for transportation. Also included in the deployment is a custom set of massive full-duplex MIMO devices, which can be scaled to 160 antennas and are operational over a wide range of frequency bands covering 470–700 MHz UHF, 2.4–2.7 GHz, and 3.5–3.8 GHz. The POWDER lab provides a few predefined profiles (packaging up the experiments with various configurations) to create its own private cloud with srsLTE, OpenAirInterface, and OpenStack-based 5G networks. The roadmap to using POWDER, description of available base-station resources, and other relevant information are available in [17].

### 13.3.2 Colosseum

Colosseum is the world's largest RF emulator designed to support the research and development of large-scale, next-generation radio network technologies in a repeatable and highly configurable RF environment. At its core, Colosseum is a specialized data center housed at Northeastern University in Burlington, Massachusetts. It has 900 TB of network attached storage, 171 high-performance servers, 256 USRP X310s (128 as communications devices, 128 as part of the channel emulator), 18 10G switches, 19 clock distribution systems, and hundreds of high-speed optical connections. Colosseum is remotely accessible by researchers and operates 24/7/365. Researchers can reserve Colosseum resources through a simple web interface. Colosseum provides users with preconfigured and ready-to-use Linux LXC containers for basic testing. Users can either use preconfigured containers like OAI 5G CN and OAI 5G RAN, or customize these containers based on their needs to develop and implement their radio codes. All these features come with a sophisticated experiment control framework, traffic generation system, real-world scenario creation, and channel reconfiguration. Several Colosseum webinars, like registration, first-time use, and container creation, are available at [18].

### 13.3.3 COSMOS

The Cloud Enhanced Open Software-Defined Mobile Wireless Testbed for the City-Scale Deployment (COSMOS) project focuses on the design, development, and deployment of an advanced wireless city-scale testbed to support real-world experimentation on next-generation wireless technologies and applications. The COSMOS testbed is deployed in West Harlem, New York City. It is intended to enable several new classes of wireless experiments not currently supported by testbeds otherwise available to the research community. Its focus is on ultrahigh-bandwidth and low-latency wireless communications with tightly coupled edge computing and emphasis on millimeter-wave (mmWave) radio communications and dynamic optical switching. It enables an entirely new class of interactive rich media applications associated with scenarios such as AR/VR for mobile users or cloud-assisted connected vehicles. The COSMOS testbed offers the following major components: (1) SDR radio nodes—100 small (both mobile and fixed), 20 medium (pole/building mount), and 9 large (rooftop installation), (2) 27 utrahigh-speed WDM optical switches for fronthaul interconnect, (3) 15 edge cloud servers colocated with large radio nodes, (4) 10 SDN switches for the backhaul network with NYSERNet high-speed connectivity, (5) 4 datacenter capacity cloud computing racks for hosting centralized testbed services and supporting both network and higher layer protocols and applications in testbed experiments, and (6) IP routers forming a private network for wide-area connectivity between the New York City deployment site and gateway to other testbeds (I2/GENI) and the internet. A COSMOS testbed overview and the user guide documentations are available at [19].

### 13.3.4 AERPAW

Aerial Experimentation and Research Platform for Advanced Wireless (AERPAW) is the first aerial wireless research platform in the United States to study the convergence of 5G technology and autonomous drones, and is led by North Carolina State University. On the AERPAW platform, a drone's communication and 5G are integrated to be mutually beneficial with increased coverage and improved signals. The AERPAW phase-1 supports USRP B205mini software-defined radios with custom unmanned aerial vehicles (UAVs) and unmanned underwater vehicles (UGVs). Open-source software such as OAI, srsLTE, and GNU Software Radio are available to run on containers for development and testbed experiment modes. Researchers generally develop experiments in a virtual environment and submit experiments for execution on the physical testbed once development is complete. AERPAW operations personnel (Ops) then execute these submitted experiments in the physical testbed environment, and the output of the experiments is available for researchers to view and analyze back in the virtual environment. The AERPAW architecture and the experimental setup are both available in the user guide document [20].

### 13.3.5 NITOS

Network Implementation Testbed Using Open-Source Platforms (NITOS) is one of the largest single-site open experimental facilities in Europe, allowing users from around the globe to take advantage of highly programmable equipment. The testbed is an integral part of a larger federation of resources, such as OneLab [21] and Fed4FIRE [22], enabling experiments that require heterogeneous resources. The major offerings from NITOS are as follows: (1) a SDR 5G testbed consisting of 10 USRPs N210, 14 USRPs B210, 4 USRPs X310, 2 USRPs N310, and 4 ExMIMO2 FPGA boards, (2) Open-Source LTE equipment, running over commodity SDR equipment, through the adoption of the OpenAirInterface platform, (3) a COTS LTE testbed, consisting of a highly programmable LTE macrocell, multiple femtocells, an experimenter configurable EPC network, and multiple user equipment (UE) devices, such as USB dongles and Android smartphones, (4) a millimeter-wave testbed, operating in the V-band (60 GHz), that supports high data-rate point-to-point setups, with beam-steering capabilities of up to 90 degrees, and (5) a cloud computing testbed, consisting of 96 cores, 286-GB RAM, and 10 TBs of hardware storage provisioned through OpenStack. The nodes are interconnected via five OpenFlow switches, sliced using the Flow Visor framework. The NITOS portal UI lists the available resources within any timeslot and researchers can remotely reserve the slices of the infrastructure that are needed to conduct an experiment. NITOS testbed hosts an OSM installation for managing the testbed components using the NFV-MANO compliant orchestrator. The link in [23] provides information about 5G virtual infrastructure provisioning over the NITOS testbed.

### 13.3.6 R2lab

Reproducible Research Laboratory (R2lab) is an open testbed located in an anechoic chamber in INRIA Sophia Antipolis, France. R2lab is part of the French FIT federation and is designed to carry out reproducible research results in wireless Wi-Fi and 4G/5G networks. Experimenters may access the facility via a portal to be used as either a sandbox or by preconfiguring configurations to be tested automatically. The 37 customizable commercial off-the-shelf wireless devices, together with USRP nodes and commercial LTE phones, overall offer a rich and flexible experimental setup. In addition, commercial phones are available for connecting to a simulated 4G network. The testbed also features advanced software such as Gnu Radio [24] and OAI, along with efficient software tools to support researchers' experimentation tasks. Both the procedures to sign up for the R2lab and example experimental scripts are available in [25].

### 13.3.7 Open Experimental Sites in 5G-EVE

The 5G European Validation Platform for Extensive Trials (5G-EVE) is one of the three 5G PPP Infrastructure projects started on July 1, 2018. The 5G-EVE goal is to interconnect four existing European sites in Greece, Spain, France, and Italy, respectively, to form a unique 5G end-to-end facility. The resulting end-to-end facility enables experimentation and validation with a full set of 5G capabilities which are Release 15 compliant. 5G-EVE technical objectives include the implementation of 3GPP Release 16 compliant technologies at all four sites and development of specific pilot activities to assess various 5G performance indicators.

#### 13.3.7.1 Open5GLab

Open5GLab at EURECOM provides experimental 5G services, including eMBB and massive machine-type communications. Open5GLab is based on fully open-source software tools and an open-architecture design. This lab is the experimental playground for OpenAirInterface and Mosaic 5g software package development. The computing resources at this site use RedHat's OpenShift 4.2 Kubernetes container platform and are used to implement radio-access, core network, and mobile-edge functions. Bare metal servers with in-lab 5G-capable radio devices are available and can be interconnected with the Kubernetes cluster. The radio infrastructure includes indoor and outdoor radio units supporting several 4G and 5G bands, specifically, band 28 (700 MHz), band n38 (2.6 GHz), band n78 (3.5 GHz), and band n258 (25 GHz). The outdoor units are interconnected with a switching fabric using 300m of fiber. It also provides remotely controllable 4G and 5G user equipment (including off-the-shelf smartphones and cellular IoT modules). The deployment framework is fully open-source and distributed through the openair-k8s GitHub [26]. Openair-k8s allows building container images for OAI 4G/5G radio access (eNB/gNB) and core networks (EPC), along with deploying these components on OpenShift or Kubernetes distribution.

#### 13.3.7.2 Plug'In

Plug'In is a platform for experimenting with 5G on top of a convergent network/IT infrastructure. The Plug'In platform is from Orange and runs on multiple OpenStack virtual machines. It provides a set of tools as an integrated development environment (IDE) in which the 5G components can be developed, experimented with, and reused. The key functionalities of these tools are:

- *AtomStore,* a central component of Plug'In. It is used to publish software modules (Atoms) that the user can develop. Atom packages are external software programs and code bits that add extra functions to the main product, thus enhancing its capabilities. AtomStore is based on the open-source private cloud nextcloud.
- *AtomGen,* a project skeleton generator to provide a homogeneous development environment. It provides a standardized structure of all Atoms to ease understanding of the different functions for the users and developers.
- *AtomDocs,* a tool to provide clean documentation for each software Atom developed. The workflow of generating the documentation is automated. This feature allows users to document their software development without worrying about formatting.
- *Toolbox* provides a starter toolkit for multiple domains that researchers can use to get involved quickly. For instance, a SDN toolbox may contain a SDN controller such as OpenDaylight or ONOS.
- *PlayGround* offers computing, networking, and storage resources to researchers to test and experiment not only on Atoms but any software. The idea of PlayGround is to enhance a reproducible research mindset and share resources.

Regarding monitoring, Plug'In uses open-source tools such as Prometheus, PCP, and Vector. Prometheus is used to monitor PlayGround metrices and both PCP and Vector are used to monitor bare-metal servers. To learn more about Plug'In, visit the link in [27].

#### 13.3.7.3 Wireless Edge Factory

Wireless Edge Factory is a cloud-native private network offering end-to-end broadband, IoT, and cellular technologies that can be deployed and run in a fully secure manner for academic and industrial researchers. Key specifications supported include:

- SDN (Openflow, OpenDaylight Controller)

- VIM: Kubernetes
- Orchestration: ONAP

Preloaded with:

- LTE EPC (OAI based) for 5G NSA support
- 5G NGC decomposed in microservices
- NB-IoT core
- LoRa Core Servers
- WLAN 802.1x Core

New releases of Wireless Edge Factory are regularly issued by b-com Research Institute of Technology and an overview of the key features is available at [28].

#### 13.3.7.4 5TONIC

The objective of the Spanish initiative called 5TONIC is to create an open global environment where researchers from industry and academia can work together on specific research and innovation projects related to 5G technologies. The key services offered in this facility include

- Provisioning of basic 4G and/or 5G connectivity to evaluate its impact on end-user application performance;
- Provisioning of measurement capabilities to evaluate key performance indicators such as throughput, latency, and/or reliability;
- Provisioning of network configuration capabilities to define the test's topology as well as other network parameters;
- Provisioning of service and application support capabilities, which enable the user to rely on the lab processing infrastructure to implement 5G use cases;
- Provisioning of adequate space and facilities for deploying the tested infrastructure of interest to the site user;
- Providing maintenance services for the infrastructure deployed by the user;
- Provisioning security services.

Some of the 5TONIC trial demonstrations include:

- Development of air technology above 6 GHz;
- Enabling dynamic operation on different frequency bands (e.g., spectrum sharing);
- Flexible radio access split;
- Massive MIMO;
- Application of mmWave technology to implement radio fronthaul and backhaul.

The laboratory supports the implementation of use cases defined by other EU projects (like 5G Vinni and 5Genesis), as well as companies that are not involved in EU-funded projects. Detailed information on 5TONIC projects and activities is available at [29].

### 13.3.8 Open Experimental Sites in 5GENESIS

The main goal of 5th Generation End-to-end Network, Experimentation, System Integration, and Showcasing (5GENESIS) is to validate 5G performance indicators for various 5G use cases in both controlled setups and large-scale events. An integrated end-to-end 5G facility is achieved by bringing together efforts and contributions from a considerable number of EU projects as well as their partners' internal R&D activities. The 5GENESIS facility implements and validates all evolutions of the 5G standard by engaging a wide diversity of technologies and chaining innovations that span all domains, achieving full-stack coverage of the 5G landscape. It enables end-to-end slicing and automation by unifying physical and virtual network elements under a common coordination and openness framework. There are five main platforms in the 5GENESIS facility. Their salient features are as follows:

- The Athens platform is a shared radio infrastructure (gNBs and small cells) that is supported by a NFV/SDN-enabled core network to showcase secure content delivery and low latency applications in large public events;
- The Malaga platform provides automated orchestration and management of various 5G NR-enabled network slices to demonstrate mission-critical services in the lab and in outdoor deployments;
- The Limassol platform combines terrestrial and satellite communications to showcase service continuity and ubiquitous access in underserved areas;

- The Surrey platform offers multiple radio access technologies in support of massive mMTC, including 5G NR and NB-IoT to experimentally investigate massive IoT services;
- The Berlin platform comprises a number of ultradense areas that are covered by various network deployments coordinated via advanced backhauling technologies to showcase immersive service provisioning.

Detailed descriptions of these platforms and their respective use cases are available in [30] under the "Platforms" tab.

### 13.3.9 Open Experimental Sites in 5G-VINNI

The EU-funded 5G Verticals Innovations Infrastructure (5G-VINNI) project aims to accelerate the uptake of 5G in Europe by providing highly advanced end-to-end facilities to validate the performance of new 5G technologies. The 5G-VINNI project comprises eight facilities: four facilities (Norway, UK, Spain, Greece) are considered as main facility sites and four other facilities (Portugal, Germany/Munich, Germany/Berlin, Luxembourg) are considered as experimental facility sites.

The main facility sites offer services to ICT-18-19-22 projects with well-defined SLAs, while the experimental facility sites offer services to researchers for focused experimentation and testing possibilities on the end-to-end model.

The resources and functional levels of the 5G-VINNI E2E facility comprise the RAN, Backhaul, Mobile Core, and Cloud Computing facilities, with a core that comes either in the form of edge or centralized clouds. These physical resources are interconnected to build dedicated logical networks that are customized to the respective telco services (e.g., eMBB, V2X, URLLC, and mMTC). Technical capabilities offered by the 5G-VINNI facilities include:

- Multiple network slices:
  - eMBB slice;
  - URLLC slice;
  - mMTC slice.
- Autonomous core by leveraging the edge resources:
  - The mobile core (both control and data plane) functionalities are pushed to the edge in the presence of backhaul connection failure to ensure self-contained operation.
- Flexible backhaul:
  - Redundancy in the backhaul is achieved via a secondary satellite link.
- Interconnection with the public cloud:

**Table 13.1**
Summary of Open-Source 5G Software Packages

| Software Packages | Access Link | Features |
|---|---|---|
| **5G Core Network Packages** | | |
| Openair-CN-5G | gitlab.eurecom.fr/oai/cn5g/oai-cn5g-fed | · Full standalone 3GPP-compliant 5G CN implementation<br>· Supports basic procedures for connection, registration, and session management<br>· Supports N2 handover, paging, and network slicing (partial support) |
| Open5GS | github.com/open5gs/open5gs | · 5GC Release 16 support using C<br>· Supports Amazon Elastic Kubernetes Service<br>· Validated on commercial 5G and RAN simulators |
| Free5GC | github.com/free5gc/free5gc | · Migrated 4G EPC to 5GC service-based architecture<br>· Supports handover and paging services<br>· Partial implementation of 5GC orchestrator, network repository function, and network slice function |
| **Evolved Packet Core Packages** | | |
| OMEC | github.com/omec-project | · 3GPP Release 13 compatibility<br>· Supports large numbers of subscribers with a high-performance DPDK-based data plane<br>· Lightweight, cost-effective deployments and supports IoT applications |
| Facebook Magma | github.com/magma/magma | · Small-scale EPC deployable in a cloud-native environment<br>· Supports orchestration, network management, and performance metric measurements<br>· Added capacity and reach by using Wi-Fi and CBRS under constrained licensed spectrum |
| srsEPC | github.com/srsran/srsRAN/tree/master/srsepc | · Lightweight implementation of complete EPC network<br>· Supports attach procedure, detach procedure, and service request procedure<br>· Supports static and dynamic IP address allocation for UE |
| **Radio Access Network Packages** | | |
| Openair-5G-RAN | gitlab.eurecom.fr/oai/openairinterface5g | · Supports 5G NSA gNB, SA gNB<br>· Supports 5G NSA and SA UE<br>· Supports RF simulator<br>· Supports CU/DU split |
| srsRAN | github.com/srsran/srsRAN | · Full-stack 4G eNodeB with 5G NSA / SA gNodeB capabilities<br>· LTE Release 10 aligned with features up to Release 15<br>· 5G NR UE modem software based support |
| O-RAN Amber | docs.o-ran-sc.org/en/amber | · O-RAN CU and DU support<br>· O&M dashboard support<br>· Interface support for RAN Intelligent Controller |

**Table 13.1** (continued)

| Software Packages | Access Link | Features |
|---|---|---|
| **Software-Defined Radio Devices** | | |
| USRP B200 | www.ettus.com/all-products/ub200-kit | · USB 3.0 interface support<br>· Frequency 70 MHz to 6 GHz<br>· Radio bandwidth up to 40 MHz<br>· Open and configurable Spartan 6 TM FPGA (XC6SLX75) |
| USRP B210 | www.ettus.com/all-products/ub210-kit | · USB 3.0 interface support<br>· Frequency 70 MHz to 6 GHz<br>· Radio bandwidth up to 40 MHz<br>· Reconfigurable Xilinx Spartan 6TM FPGA for Advanced Users (XC6S150) |
| X300 | www.ettus.com/all-products/x300-kit | · 10G Ethernet PCIe interface support<br>· Frequency up to 6 GHz (based on daughter board)<br>· Radio bandwidth up to 80 MHz<br>· Large customizable Xilinx Kintex-7TM FPGA (XC7K325T) for high-performance DSP |
| N300 | kb.ettus.com/N300/N310 | · 2 x 10G Ethernet interface<br>· Frequency 10 MHz to 6 GHz<br>· Radio bandwidth up to 100 MHz<br>· Xilinx Zynq-7035TM FPGA SoC |
| Blade RF 2.0 micro | www.nuand.com/bladerf-2-0-micro | · USB 3.0 interface support<br>· Frequency 47 MHz to 6 GHz<br>· Radio bandwidth up to 40 MHz |
| LimeSDR-mini | limemicro.com/products/boards/limesdr-mini/ | · USB 3.0 interface support<br>· Radio bandwidth up to 50 MHz |
| **Control and Orchestration Packages** | | |
| OAI Mosaic 5G | gitlab.eurecom.fr/mosaic5g | · Flexible RIC support<br>· Interface with OAI Radio Stack for monitoring and slicing functionality<br>· Flex RIC compatible with O-RAN RIC |
| Akraino | github.com/akraino-edge-stack | · Fully integrated edge infrastructure solution<br>· Supports high-availability cloud services for edge computing applications<br>· Scale edge cloud services quickly with Kubernetes orchestration |

- Network functions are hosted in or extended to the public cloud.
- Interconnection with other 5G-VINNI facilities;
- Interconnection with non-5G-VINNI facilities.

## 13.4 Summary

In this chapter we provided an overview of open-source core network, radio-access network, and edge computing software packages that can be readily

**Table 13.2**
Summary of 5G Experimental Sites

| Experimental Site | Access Link | Salient Features |
|---|---|---|
| POWDER | https://powderwireless.net | · Simulated and controlled RAN environment<br>Over-the-air environment<br>Massive MIMO base station<br>Vehicle-based mobile endpoint<br>Portable mobile endpoint |
| Colosseum | https://colosseumneu.freshdesk.com | · Beamforming/massive MIMO infrastructure<br>· Fusion of artificial intelligence and the Internet of Things<br>· Realistic RAN deployment<br>· Spectrum sharing |
| COSMOS | https://cosmos-lab.org | · Massive MIMO/mmWave access<br>· Integrated SDR support<br>· Optical switching network<br>· Smart city intersection use cases |
| AERPAW | https://aerpaw.org | · Drones and 5G integration<br>· Autonomous driving feature test<br>· Aerial base-station support |
| NITOS | https://5ginfire.eu/nitos | · Millimeter-wave testbed<br>· SDR-based 5G testbed<br>· Open-source and commercial-off-the-shelf LTE testbed |
| R2LAB | https://r2lab.inria.fr | · Integrated SDR devices<br>· Commercial Huawei LTE sticks<br>· Commercial 4G/5G Sub-6 GHz module |
| Open5GLab | http://open5glab.eurecom.fr | · Open testbed based on OpenAirInterface and Mosaic 5G platforms<br>· Enhanced mobile broadband feature<br>· Massive machine-type communication feature |
| Plug'In | https://atomdocs.pluginthefuture.eu | · 5G experimentation on converged network infrastructure<br>· Provides sandbox (tools) for instantiating and testing 5G components |
| Wireless Edge Factory | https://b-com.com | · 4G, 5G, NB-IoT support<br>· Network slice-based isolated services<br>· Augmented reality, virtual reality<br>· Artificial intelligence |
| 5TONIC | https://www.5tonic.org | · 5G virtual software network—IoT<br>· 5G wireless system<br>· 5G-crosshaul data plane<br>· Air technology above 6-GHz band |
| 5GENESIS | https://5genesis.eu | · Five facility sites<br>· Edge computing enabled shared radio infrastructure<br>· Orchestration and management of network slices<br>· Integrated terrestrial and satellite radio communication<br>· 5G NR and NB-IoT support |
| 5G-VINNI | https://www.5g-vinni.eu | · Nine facility sites<br>· 5G NR support<br>· Network slicing<br>· MANO and service orchestration<br>· Satellite installation and operational |

used to build PoC experimental 5G networks. These open-source projects and their key technical features are summarized in Table 13.1. We also provided an overview of existing experimental facilities in both the United States (U.S.) and European Union (EU) that offer remote access to various experimental infrastructures that may be used to prototype 3GPP radio and core network technologies. These facilities and their salient features are summarized in Table 13.2. A more detailed survey of software tools, deployment and testing methodologies, and computing platforms used in major U.S. and EU platforms can be found in [31]. By their evolutionary nature these community platforms and initiatives are expected to continue to evolve, grow, and adapt to address the future technical challenges that engineers will face in deploying 5G and 6G solutions. The number of these initiatives and their relative sizes are indicative of the volume of investments that are going into the exciting mission of identifying and addressing the numerous technical problems of designing these future mobile networks.

## References

[1] https://openairinterface.org/oai-5g-core-network-project/.

[2] https://open5gs.org/open5gs/docs/.

[3] https://opennetworking.org/comac/.

[4] https://docs.srsran.com/en/latest/usermanuals/source/srsepc/source/1_epc_intro.html.

[5] https://openairinterface.org/oai-5g-ran-project/.

[6] https://openairinterface.org/oai-code-2/.

[7] https://www.srs.io, December, 2022.

[8] https://docs.srsran.com/en/latest/feature_list.html.

[9] https://wiki.o-ran-sc.org/display/TOC/Project+Readiness+for+Amber+Release.

[10] www.ettus.com.

[11] www.ni.com.

[12] www.nuand.com.

[13] https://bubbleran.com/media/blog-post.2/.

[14] https://openairinterface.org/mosaic5g/, December 2022.

[15] https://wiki.akraino.org/display/AK/Akraino+Edge+Stack.

[16] https://www.fieldengineer.com/blogs/what-is-dark-fiber.

[17] https://docs.powderwireless.net/.

[18] https://colosseumneu.freshdesk.com/support/solutions/articles/61000284565-colosseum-webinars.

[19] https://wiki.cosmos-lab.org/wiki/WikiStart#COSMOSTestbedOverview.

[20] https://sites.google.com/ncsu.edu/aerpaw-wiki/aerpaw-user-manual.

[21] https://onelab.exelonpowerlabs.com/.

[22] https://www.fed4fire.eu/, December 2022.

[23] https://5ginfire.eu/nitos/, December 2022.

[24] https://www.embedded.com/understanding-software-defined-radios-and-networks-in-5g-architectures/.

[25] https://r2lab.inria.fr/tutorial.md.

[26] https://github.com/OPENAIRINTERFACE/openair-k8s.

[27] https://pluginthefuture.eu/.

[28] https://mycfia.com/uploads/products/brochure/985d773c-6e67.pdf.

[29] https://www.5tonic.org/projects-and-activities/.

[30] https://5genesis.eu/.

[31] https://www.advancedwireless.eu/deliverables/.

# 14

# Summary of Virtualization of 5G and Beyond

## 14.1 Where It All Began

Martin Cooper invented the cell phone [1] while at Motorola and placed what is now the famous call on April 3, 1973, to AT&T Bell Labs head of research Joel Engel. There has been no stopping the progress of the mobile network even though it took nearly 9 years for the Federal Communications Commission (FCC) in the United States to finally authorize commercial cellular service in 1982. Interestingly, this is the same year in which AT&T was broken up by the U.S. government under the AT&T divestiture decree. Five years would pass before the FCC would allocate additional spectrum in 1987 to address the customer demand that had outstripped the initial design limitations. Some may find it interesting that AT&T Bell Labs at the time showed limited interest in this new technology, and AT&T would enter and exit the mobility business in the early years. The company brand was reconstituted into at&t (hence the change from capital letters to lower case) when Southwestern Bell acquired AT&T in 2005, and shortly thereafter added Cingular Wireless, (which AT&T has sold off previously). This was a case of getting the band back together under the old name but with a new lead singer, the Southwestern Bell leadership, using the lower case at&t name. While all this was going on the standard-bearer of research for the industry, Bell Labs, changed hands several times and is now part of Nokia.

The introduction of the 3GPP standards body brought with them a global standard that unified what were otherwise incompatible technologies and unleased an ecosystem to drive a global industry forward. In 2007, Apple's introduction of the iPhone based on the 2G standards drove a rush to increase network capacity that ultimately led to the 2012 NFV vision paper addressing an emerging concern that the operators would not be able to sustain the investments necessary to keep up with the rapidly growing demands on their network.

Following the model laid out a decade earlier by large cloud providers, the separation of hardware from software provided network operators with a model that would allow them to keep pace with the growth demands on their networks. This disaggregation, however, was more than just the separation of hardware from software—it complicated system integration and maintenance of the "single pane of glass" management that the telecom industry was familiar with and had spent decades fine-tuning to streamline operations and reduce operating costs. In the decade that followed the NFV vision paper, the operators found that virtualization of the core signaling components was straightforward and the standards bodies applying the CUPS gave them a new model to use in building and expanding their networks, including transforming their operational practices.

However, many operators have discovered over the past decade that while they are very good, if not excellent, at operating a network, they have been ill prepared to take on the responsibility of acting as the system integrator for their revenue-generating network. The disaggregation of the network functions from previously being supplied on purpose-built systems brought with them responsibility for support, interoperability, and defect identification and resolution. Many leading communication service providers today have found that changing the support model where a system in their network could be supported from at least three separate vendors—one for the hardware, one (or more) for the middleware, and one for the workload—has placed additional triage responsibility on the CSP team. These teams in some cases may lack the technical capabilities to correctly identify the defects in the various layers, or even worse, may at times be unable to identify responsibility when there is no clear demarcation for the defect, such as if the defect is not the result of one vendor's shortcoming but due to design choices made by two or more vendors. This is a common occurrence today, and this is beginning to be recognized by CSPs as the hard part of the transformation.

One leading CSP in the United States may have a start at addressing this issue, while others may find that taking the same or similar path may aid in closing the above gap. For example, AT&T has transitioned the operational support of their 5G core to Microsoft. This transition, when combined with the Microsoft acquisition of Metaswitch and Affirmed in 2020 may be viewed as the beginning of the hyperscalers bringing their own view of how DevOps can and should be implemented in the new disaggregated telecom network model.

Other CSPs are experimenting with improving their virtualization capabilities or looking to suppliers further down the ecosystem chain to step up and provide the triaging of issues and support to resolve the integration issues. This model may not survive as it places a second layer of responsibility for defect resolution.

The model being applied by new entrants in green-field deployments have the advantage of a sheet of paper (i.e., no legacy networks/operations to adapt or migrate). An example is Rakuten Symphony. Rakuten acquired their RAN vendor Altiostar and deployed a team to develop a new product model to replicate the network design at scale found in Rakuten's network. Rakuten is now offering their network software and design as a product to other carriers although not in their competitive footprint. This model is somewhat reminiscent of the very old model of the largest carriers, such as the original AT&T who could sell their PSTN switches to carriers providing them with a proven hardware system and operating model in a "copy exactly" package.

A recent report from SLT Partners coined the term "hyperscaler economics" [2]. The authors provide two ways of visualizing the concept of hyperscaler economics. The first is a linear cost base when facing an exponential growing capability (we would translate this to capacity in the network), and the second is to visualize the same concept to linearly grow the capability while realizing an exponentially decreasing cost model. These two visualizations are equivalent.

This report is the result of input from the leaders at 14 CSPs around the globe, 7 of whom are operating in a single market as might be found in the United States. This group includes AT&T, Verizon, and possibly T-Mobile. Others are operating with a group level model, including Vodafone along with possibly three others. And finally, two or three are operating with an OpCo model. This report clearly points to the transformation and challenges that are being played out in the global network today as the operators transform the way they build and operate their networks. Many of the respondents, but by no means all, converge on several common themes including the need to transform the talent and operational structure and possibly even the culture of their teams. They recognize that this is a journey will take time and energy. Some of the next-wave operators have expressed concern over the challenge of attracting the talent necessary for this cultural transformation as the disaggregated model places new technical responsibility on the support staff within the CSP, whereas in the days of purpose-built systems this responsibility resided with a sole provider of the network element. In addition, there is a general agreement that the terms of cloud native, and disaggregation, at least among the executives polled at these 14 operators, were often at times confusing and possibly not clear. What was clear though, was the need to transform the economic model used to build and operate the network and to adapt to a business model that permitted innovation to be developed within the CSP. Extrapolating some of the discussion from the 14 CSPs in this report, it is clear that there are innovative CSPs leading the way who will drive the technology forward. In the near term, adopters will follow a similar path to these leaders, while other CSPs in either emerging markets or for some other reasons will follow sometime later. The question is not if the transformation is happening, but rather at what rate, and what the next steps may look like once the technology is operationalized. For example, will the CSPs ultimately achieve a software industry CI/CD model, or will they create a CI/CD model that resembles a vendor-provided model that is easier to adopt within their organizations and are compatible with their legacy infrastructure? Nevertheless, this report serves to validate the discussion in this book of the journey that is the transformation of the 5G and Beyond network, from the edge to the core.

Another recent report provides analysis and a preliminary recommendation from the European Union (EU). This report by the Axon Partners Group [3] for European Telecommunications Network Operators Association (ETNO) discusses the impact and investment the European operators have been making in their very high-capacity networks (VHCNs) over the past decade, including fiber to the home (also referred to as fiber to the premises) and 5G networks. While some of the recommendations and the conclusion of this report will likely be contested, and may even eventually be overturned, it represents a current view from the European Union and may reflect the pressure on the network operators in other jurisdictions.

The report claims that the community of operators have invested over 500B Euros during this time. While the report does not separate out the 5G investment, the magnitude of this number should not be ignored. It is a vast sum of money, and the driver for this expansion in capacity has been due to the volume of OTT traffic, or as others consider it, through-the-network traffic. The economic investment in the network by providers of OTT services is negligible compared with the network investments, and the operators have seen little economic gain from this investment. The report further points out that neither the operators nor the regulators are in any position to bring these hyperscalers to the negotiation table to discuss commercial terms. This may require resolution in the legal system, rather than the technical field, and is likely to vary from jurisdiction to jurisdiction.

The report goes on to offer regulatory remedy and suggests that the largest hyperscalers be compelled through regulation to contribute a fair and proportionate share to cover the "costs of public goods, services and infrastructures" [4].

While the analysis of this report is focused on the EU members and the possible ways that the hyperscalers can be brought to the investment table, it is by no means exclusively a problem for Europe. Nearly every jurisdiction is struggling with these same economic and competitive pressures.

One of the innovative features of 5G is the ability of the 5G RAN to use portions of the 4G core, which was discussed earlier and is known as NSA 5G, while pure 5G deployment is known as SA 5G. A report released from STL Partners [5] outlines their research on global network CSPs core plans and reports that 212 (71%) of the polled carriers planned to remain on a NSA core for 5G at least into 2023. In addition, 66 respondents (22%) of those responding, are already or will be on SA by 2023, with the remaining 19 respondents (7%) on a converged NSA/SA core. This implies that a large number of operators have chosen to remain on the 4G core while upgrading the RAN to the 5G standards. As discussed, there may be limits to the full set of 5G capabilities in these configurations. While the 5G RAN will provide the greater bandwidth, possibly up to 10 Gbps, provided the channel allocation is at the higher bands, there may be limits on the ability to achieve either the density or latency improvements implied in mMTC and uRLLC.

## 14.2 New Markets

One theme that has been discussed several times is the need for the telecom operators to recover their enormous investment in expanding their network into the 5G space. Clearly, there is limited opportunity in today's highly competitive market to increase the monthly subscription fee of the individual consumer ARPU. Nor is there much opportunity to expand the number of consumers in most developed markets. In addition, there is limited opportunity, until such time as regulators choose to act, for the telecom operators to realize revenue from OTT services or through the network provided by the hyperscalers and others. This leaves one clear potential market—to expand into the enterprise with features and services beyond pure connectivity. Here too, many of the existing operators are facing challenges. Many of the current challenges are less technical and more related to current business practices such as institutional sales habits of selling connectivity rather than higher-value services. The emergence of Industry 4.0, private wireless networking, and aggressive pressure from the hyperscalers and from some network equipment manufactures may be sufficient motivation for many carriers to accelerate their transformation into the commercial space and offer 5G solutions. 5G offers the advantages of network slicing where the operators can implement customized and virtually dedicated private networks where either regulatory or business practices require security or capacity that is tuned to the user. A common trend is that the next generation of technology, in this case 6G, benefits more from these advances as both the timeline for the maturation and adoption in the market can be significant.

## 14.3 6G Is on the Horizon

6G is being discussed both in the working sessions of 3GPP today and with various early technology developers. While there is no agreement yet on 6G features, several trends are emerging. One is to provide even greater bandwidth, up to 5× the current upper limit of 20 Gb/s to 100 Gb/s download. As discussed earlier, this implies that more radio spectrum will need to be allocated by the various regulators. In addition, it will be necessary to bring additional customers, enterprise, and others to the market to help offset the investment costs the telcos will face. Cloud native architectures and edge compute will continue to be at the forefront of these discussions as the next generation is architected. Today, some of the RF test equipment providers are starting to offer the equipment that will be necessary to test prototype RRUs in the anticipated new spectrum bands. It is reasonable to assume that when 6G begins to mature, we will see both NSA and SA models as we have seen in 5G. The primary driver here is to allow the operators to introduce 6G radios while maintaining their preexisting 5G core networks until such time as the market allows for the additional investment in the core to meet the 6G design goals [6].

Some of the features in 6G will address any shortcomings found in 5G through practice and scale, and other features will likely be innovative new use cases. One interesting area may be the evolution of regulation for example allowing hyperscalers to bid on spectrum usage. This could possibly alter the landscape in ways never imagined by the traditional operators. The hyperscalers have shown unprecedented ability to innovate and will likely find ways to close the above-mentioned network investment barriers.

While there is as yet no fixed vision or driver for the 6G specification work underway today, what we do know is that the spectrum allocation is likely to be well above the millimeter-wave bands used for 5G, possibly in the terahertz range. The continuation of wider channels (more bandwidth), improved latency reductions, and the introduction of AI to manage the network performance and power demands are all in the ongoing discussions. The remaining missing piece remains: How are the operators going to monetize future investments?

While our primary focus in this book has been on virtualization for the 5G mobile network and beyond, there was a development announced by Samsung in the summer of 2022 that is worth mentioning here [7].

While many MVNOs have already or plan to decommission their 3G networks soon, Samsung is building a new vRAN for 2G. While the commercial viability may be debated, the technical validation of virtualization is borne out by this development. One may rightly ask, why would anyone (vendor or operator) be investing in 2G now? We offer a few observations: first, the availability of "new" 2G equipment from the legacy build-out is no longer an option. Last-time buys of this equipment have long expired, and the stock of the custom ASICs that were part of that solution have long been exhausted or relegated to the recycle bin. This does not happen with software of course, and the availability of programmable device like FPGAs in the RRUs makes it possible to program the FPGA to the 2G RAN requirements. In addition, the mobile core to support 2G can be completely realized on modern high-volume servers without any additional acceleration requirements. Now, we may ask, how is this a viable business plan, as the need for a 2G handset may also be difficult to justify. There are two markets: one is the very underserved and very poor regions of the global community that may benefit from a low data rate and very low-cost handset. These developing nations may have in total over 1B consumers who could gain access to capital markets and other resources with a low data rate connection device. In addition, the physical range of 2G networks (due to the band of operation primarily) is greater than that of 4G or 5G networks. Thus, fewer RANs are required to span a large geography. Time will be the judge if this is a good business decision, yet there is another potential market opportunity—the legacy IOT market where industry and enterprises have a large installed base of devices and infrastructure that they may want to expand. Legacy operators have expressed their desire or intent to exit this market, but the possibility of operating a private network using this technology may be attractive to a large enough community to make this a winning business opportunity. Another possible opportunity lies in the electricity utility sector where mesh networks operating in unlicensed spectrum currently support many millions of smart meters for energy monitoring and flexible tariffs. This global industry may find a fit for private spectrum (regulatory approval would be required) to improve the coverage and reduce their overall cost. Virtual 2G in this market may play well because of the very low power demands of the UE, and the utility has to cover this power budget, and the lower operating cost with a much greater reach. Again, time will be the judge and we have likely over speculated on a point or two, and also freely acknowledge that we may have underestimated the market demand for a virtualized 2G network solution.

## 14.4 Summary of Some Key Factors

There remain three key economic factors. The first is the cost of the solution (purchasing the equipment, licensing the software, and installing all the physical devices into the network), the second is operational staffing and support costs, and the third is the cost of power to operate the system. By some rough estimates these costs are more or less equal over the lifecycle of these solutions in a modern CSP network. The vision outlined in the seminal NFV white paper was that virtualization and the replacement of vendor-specific hardware with high-volume industry standard servers would enable disaggregation of the network elements and provide a mechanism to address each of these three areas. With over a decade now past, it is fair to ask how the industry has done in achieving the goals outlined in that paper.

### 14.4.1 A Cloudy Crystal Ball

One key learning in the disaggregated model, albeit obvious, is that ecosystem partners will always operate in their own long-term best interests. For example, one middleware OS provider ceased to provide the unlicensed version of their operating system that caused some of the smaller ISVs to pivot to a different operating system, incurring increased product development costs and leading to schedule delays. Some vendors of accelerator cards (for example the FPGA cards used by some ISVs for FEC acceleration) chose to prematurely declare their products obsolete, forcing ISVs to replan at short notice and sustain a schedule impact. These events are not likely to occur with a purpose-built solution as one vendor usually has control of all the elements on which their solution depends.

A collection of approaches may be considered to mitigate the possible concerns above. The first is a contractual approach, entering into binding agreements with penalties for the support of the product over the service lifetime. While the timeline for introduction into the network may span from a few years to five or more years, the operators will have an expectation for support (both for the hardware and software) for the operational life of the technology in the network. In telecoms, the scale and geographic span of network deployment means that the technology will be operational for 10 to 15 years or more. In the case of 2G, that span would be over two decades, and for the PSTN over 30 years.

A second approach is to base the solution on the most current offerings from a downstream ecosystem partner. While this may help in simplifying the choice of hardware and any acceleration functions, it is no guarantee because availability of specific servers and specific CPUs may be much shorter than the expected operational service life. In this case the virtualization with good inventory control may help avoid unexpected supply interruption. Many CSPs will insist on an RMA model for installed equipment within their remit and this requires spares to be available on a very short turnaround, often within 24 hours or less. This is clearly necessary for the equipment located remotely at or near cell towers such as the RRU and DU.

The third approach is more difficult, as the ISV will need to invest in more development effort to build a product that is truly cloud-native to be more portable to different platforms, even for elements close to, or including, the RRU. This model would abstract any hardware acceleration functions and support a case where CPU cycles could be used, possibly reducing processing capacity, increasing power, or introducing delay, at the expense of true portability.

A final observation is that it is still the common expectation of many CSPs that there is a single entity (not them) who is responsible for the overall system integration work including resolution of defects or operational anomalies that may be uncovered.

## 14.5 Conclusion

Disaggregation through virtualization is clearly the way of the future for telecom network design and implementation. The economics require it and the hyperscalers have shown the way. Many of the issues that remain are more business related than technical, although technical issues will continue to be raised that need to be resolved. What remains to be seen is whether the approaches to virtualization will provide true multivendor options across network or within network elements. Some of this is also business related as the largest vendors would need to decide to disaggregate their product offerings and offer an adequate level of support to fully disaggregated deployments. Finally, there is evidence [8] that market segmentation based on disaggregation may be developing in the vRAN space.

### 14.5.1 Possible Research Areas

The cloud technology-centric virtualization approach advocated in the original NFV paper arose through the R&D efforts of the large carriers who coauthored the paper. Going forward, the R&D efforts of the hyperscalers will be as important, if not more important, than that of the carriers themselves who have progressively reduced their R&D expenditure, instead becoming reliant on the R&D insights coming from their large vendor partners whose agendas may not necessarily be aligned to protect their self-interest. Reducing power consumption while increasing performance to handle larger network capacities and more complex application workloads remain critical aspects of any large-scale commercial deployment in the global telecom operational network. While time to market for new services is also a central concern, the ability to operate these systems to meet increasing demands for sustainability will require the ability to utilize the deep analytics capabilities available on modern CPUs in order to make the intelligent and timely decisions on the investments needed to keep pace with the ever-changing demands of the consumer and business while remaining competitive amongst an expanding landscape of players.

## References

[1] Loeffler, J., "The History Behind the Invention of the First Cell Phone," *Interesting Engineering,* January 24, 2021, https://interestingengineering.com/innovation/the-history-behind-the-invention-of-the-first-cell-phone.

[2] Luk, Y., "Pursuing Hyperscale Economics: What, Why and How Telcos Can Do It," STL Partners, https://stlpartners.com/wp-content/documents/reports/Pursuing%20hyperscale%20economics.pdf.

[3] "Europe's Internet Ecosystem: Socioeconomic Benefits of a Fairer Balance Between Tech Giants and Telecom Operators," Axon Partners Group, May 11, 2022, https://www.axonpartnersgroup.com/europes-internet-ecosystem-socio-economic-benefits-of-a-fairer-balance-between-tech-giants-and-telecom-operators/.

[4] European Draft Declaration on Digital Rights and Principles for the Digital Decade, COM (2022) 28 final, 26.1.22, European Union, https://digital-strategy.ec.europa.eu/en/library/declaration-european-digital-rights-and-principles.

[5] "5G Standalone (SA) Core: Why and How Telecos Should Keep Going," STL Partners, September 2022, https://stlpartners.com/research/5g-standalone-core-why-and-how-telcos-should-keep-going/.

[6] Meyer, D., "6G Needs Edge, Cloud Cooperation, ABI States," *SDX Central,* September 22, 2022, https://www.sdxcentral.com/articles/analysis/6g-needs-edge-cloud-cooperation-abi-states/2022/09/.

[7] Meyer, D., "Samsung Preps 2G vRAN Option," *SDX Central,* June 29, 2022, https://www.sdxcentral.com/articles/news/samsung-preps-2g-vran-option/2022/06/.

[8] Morris, I., "Intel Risks Losing Arm Wrestle as Open RAN Splits into Rival Camps," *Light Reading,* November 9, 2022, https://www.lightreading.com/open-ran/intel-risks-losing-arm-wrestle-as-open-ran-splits-into-rival-camps/d/d-id/781641.

# Glossary of Acronyms and Common Terms

We have compiled a list of some of the alphabet soup found in the industry, including many used in this work. We have attempted to provide an expansion of these acronyms, including three-letter acronyms (TLAs), in line at first use/appearance in this book; our apologies if we missed a few. But it is also good to have these in a single place, right here. We've expanded the acronym, abbreviation, or TLA, and provide a brief definition. A fast search on the web may also serve the curious on first encounter of an unknown term.

An initial guess is there may be several hundred additional terms that have been abbreviated in the CSP industry, so many are not given here; it is not our intent to try and list them all in this book, nor to provide a complete definition to each term. That work has proven to be a separate book on its own. Decades ago, Harry Newton built a business publishing *Newton's Telecom Dictionary*, and at last check some copies are available online. Let's get started.

**3GPP** 3rd Generation Partnership Project, a quasi-standards body that defines the mobile standards

**5G** 5th Generation mobile network protocol and standards as defined by 3GPP

**5GC** 5G Core network, the functional elements comprising the user and control plane of 5G network

**5G-PPP** 5G-Public Private Partnership; see 3GPP

**AI** artificial intelligence, a practice and theory of computer science to solve nonalgorithmic problems

**AMF** access and mobility management function, a control function in the 5G core modules

**API** application programming interface

**ARPU** average revenue per user, the average monthly fee subscribers are charged for mobile access

**ARQ** Automatic Retransmission Request, an error correction protocol using retransmission of packets

**B5G** Beyond 5G, referring to 3GPP standards in the post-5G era (in 2028 and later)

**BBU** baseband (processing) unit, an element of a radio where all data are processed at a baseband frequency

**BFP** block floating point, the method used to provide arithmetic approaching a floating point by using the fixed-point processor

**CA** certification authority, a trusted authority that issues security certificates

**CapEx** capital expenditures, a type of expense usually used to add capacity or resources to a business

**CI/CD** continuous integration/continuous delivery, a software development and deployment strategy

**CIS** Center for Internet Security

**CMP** Certificate Management Protocol, an IETF (RFC 4210) standard to manage certificates

**CNF** cloud-native network function, typically a container-based workload in software

**CoMP** coordinated multipoint combines antennas from various small cells to create more spatial dimensions, thereby increasing system capacity

**CommSP or CoSP** communication service provider, sometimes CSP; see CSP telco

**CP** control plane, a path and service of control and signaling data in the mobile network

**CUPS** control and user plane separation, a theory that separates user data flows from control

**CPRI** Common Public Radio Interface, a vendor standard for the fronthaul in the RAN

**C-RAN** centralized radio access network

**CRI-O** container runtime interface for OCI-compatible runtimes

**CRMT** core root of trust measurement

**COTS** commercial off-the-shelf, a server available, for example, from HPE, Dell, QCT, or Advantech

**CSP** communication service provider, a telco (e.g., AT&T, Vodafone, Reliance Jio, China Mobile)

**CU** central unit, the separation in the 5G RAN between the DU and 5G Core elements

**CU-CP** central unit control plane, the signaling and control functions in the CU

**CU-UP** central unit user plane, the processing functions of the user data in the CU

**CUPS** control and user plane separation; 4G and later architectural design separating data and control

**CUS** control, user, & synchronization

**DL** downlink, traffic that is flowing from the direction of the core toward the end user

**DOS, also DoS** denial of service, a hostile attack intended to overwhelm a system or network

**DDOS** distributed denial of service, an advanced DOS attack originating from many sources

**DSS** Dynamic Spectrum Sharing technology to push 4G and 5G simultaneously

**DTLS** Datagram Transport Layer Security

**DU** distributed unit, the node in the 5G network between the RU and the CU, usually close to the RU

**EMS** element management system, a software function providing operational information, control, and statistical data on a system and solution in the network

**EN-DC** LTE-NR Dual Connectivity

**EPC** Evolution Packet Core, the 4G core for control and data plane functions as specified by 3GPP

**EPC-NAS** EPC Non-Access Stratum, the protocol to manage the signaling sessions with the UE as it moves

**ETSI** European Telecommunications Standards Institute, a standards body founded in Europe

**EST** Enrollment over Secure Transport

**eCPRI** Enhanced Common Public Radio Interface, an open standard of CPRI found in oRAN

**eMBB** Enhanced Mobile BroadBand, a concept within 5G providing 10× the 4G bandwidth

**eNB** eNodeB, the proper name of the base station for LTE/4G RAN

**gNB** gNodeB, the proper name of the base station for 5G RAN

**en-gNB** gNB with only secondary node functionalities

**FIPS** Federal Information Processing Standards, a standards body

**FQDN** fully qualified domain name—an entity's complete address on the internet

**GPRS** General Packet Radio Service within GSM, UMTS, LTE, and 5G NR

**GSM** Global System for Mobile Communications, a set of standards for radio access

**GSMA** Global System for Mobile Communications Association, a standards body

**GTP** GPRS Tunneling Protocol, to establish a channel between GPRS supported nodes

**HARQ** Hybrid Automatic Retransmission Request, an enhanced error correction protocol

**HSM** hardware security module

**HTTP** HyperText Transfer Protocol, a protocol to transfer information between servers and browsers

**ICAM** identity, credential, and access management

**IDE** Integrated Development Environment, a platform to facilitate software development

**IETF** Internet Engineering Task Force, a standards body focusing on Internet Protocols

**IMT** international mobile telecommunications, a term sometimes used to refer to the global CSP network

**IQ** in quadrature, or two signals used in coherent modulation schemes

**ISV** independent software vendor, a commercial entity that sells or license software rather than compute systems (hardware) that may include their software

**ITU** International Telecommunication Union, an agency of the United Nations

**LLS** Lower Layer Split, location of separations in the protocols of 5G low in the stack

**LUKS** Linux Unified Key Setup

**MAC** Mandatory Access Control

**MAC** Media Access Control (address), the 48-bit L2 address found in IP packets

**MANO** management and network orchestration, a key set of functional elements defined in the 2012 NFV whitepaper

**MEC** Multiaccess Edge Computing, or Mobile Edge Computing, a standard and also a term used in edge network elements

**MIMO** multiple-input multiple-output, in 4G and beyond to send and receive on multiple antennas

**MITM** man-in-the-middle, an attack vector where a bad actor is able to intercept and possibly modify a secure communication from within the network

**ML** machine learning, a specific branch of AI

**MN** master node

**MNO** mobile network operator, another term for CSPs holding licenses and operating radio spectrum

**Multi-RAT** Multi-Radio Access Technology, in support of dual connectivity

**mMTC** massive machine-type communications, a 5G concept allowing 10×+ the connections over 4G

**NAS** Non-Access Stratum protocol of the 4G EPC

**NAS** network attached storage

**NDS** network domain security

**NESAS** Network Equipment Security Assurance Scheme

**NEP** network equipment provider, also known as TEM; see TEM

**NF** network function

**NFV** network function virtualization, a term coined in the 2012 white paper, separating hardware from software

**NFVi** network function virtualization infrastructure, the middleware in a NFV node

**NG-RAN** Next Generation RAN, the 3GPP radio access network for 5G

**NGEN-DC** dual connectivity configuration utilizing the 5GC

**NIST** National Institute of Standards and Technology, a U.S. standards body

**NR** New Radio, often used in discussing the 5G RAN

**NR-RIC** near real time RIC, oRAN introduces a radio intelligent controller to improve RAN performance

**NSA** Non-Standalone 5G Architecture, connecting the 5G RAN to a 4G Core

**Near-RT RIC** near-real-time RIC, oRAN real time portion of the RIC, see RT-RIC

**ng-eNB** evolved LTE base station

**non-RT RIC** non-real-time RIC, oRAN, NR-RIC replied in < 1s, often <1 ms, non-RT is > 1s

**OCI** Open Container Initiative

**O-CU** O-RAN central unit

**ODM** original design manufacture, an entity that designs and builds servers to others design specs

**O-DU** O-RAN distributed unit

**OEM** original equipment manufacturer, an entity that builds servers to their own design specs

**O-RAN** Open Radio Access Network, the Open RAN design specified by the oRAN alliance

**O-RANa** Open-Radio Access Network alliance, see www.o-ran.org

**oRAN** see O-RAN, a different way to write O-RAN

**O-RU** O-RAN Radio Unit, the Open RAN RU

**OpEx** operational expenses, expenses related to running the business, often labor, overhead, electric power consumption

**OTT** over the top, normally referring to a cloud or hyperscaler consuming mass amounts of data through the CSP network providing services (e.g., often video)

**OxM** original equipment and design (OEM, ODM) manufactures, builders of hardware

**PDCP** Packet Data Convergence Protocol

**PDCP-C** Packet Data Convergence Protocol–control plane

**PDCP-U** Packet Data Convergence Protocol–user plane

**PDU** protocol data unit, a block of information transferred over a network in the OSI model

**PHY** physical layer, the lower layer in the OSI model where raw bits are transmitted and received

**PoC** proof-of-concept designs are often used before defining standards and developing products

**PNF** physical network function

**PTP** ITU Precision Time Protocol with telecom profile in support of time synchronization in the network

**QFI** QoS flow identifier in both downlink and uplink packets

**QoS** quality of service

**RAN** radio access network, a major portion of the wireless network using RF to connect users

**RAT** radio access technologies

**RBAC** role-based access control

**RF** radio frequency, a portion of the electromagnetic spectrum suitable for use in telecommunications

**RIC** Radio Intelligent Controller, oRAN defined radio control element into the 5G protocol stack; in chapters and in online search, RIC is RAN Intelligent Controller

**RLC** radio link control

**rrc** radio resource control

**RRH** remote radio head; see RRU

**RT-RIC** Real-Time Radio Intelligent Controller

**RRM** radio resource management

**RRU** remote radio unit, also RRH, the RF processing unit, usually near the antennas

**RU** radio unit, the portion of a 5G radio that is analog and usually high power, connects to DU

**SA** standalone 5G architecture, in 5G where the RAN and core are both 5G-compliant

**SAST** static application security testing

**SCTP** Stream Control Transmission Protocol, an IETF standard that improves on TCP

**SCRM** supply chain risk management

**SDAP** Service Data Adaptation Protocol, for flow-based quality-of-service capabilities

**SDN** software-defined networking, the disaggregation of control and data for routing and switching

**SDLC** software development life cycle, the timeframe a software product is supported

**SDR** software-defined radio—boards implementing radio frequency circuitry

**SGX** Software Guard eXtension, SGX offers hardware-based memory encryption that isolates specific application code and data in memory

**SHVS** SGX Host Verification Service, the SHVS collects the SGX information of all compute nodes in a data center

**SIEM** security information and event management

**SLA** service-level agreement, a business relationship with contractual defined performance obligations and may include penalties

**SMO** service management and orchestration

**SN** secondary node, a second node

**SS7** Signaling System 7, a legacy telecommunications standard for signaling in the network

**SSH** Secure Shell, a Linux (UNIX) Command Line tool that uses encryption

**STG** security task group

**SUCI** Subscription Concealed Identifier

**TCO** total cost of ownership, the sum of all costs during the life of a service

**TEM** telecommunications equipment provider, a vendor specializing in building equipment that is consumed mainly by CSPs

**TLS** Transport Layer Security, a cryptographical protocol

**TPM** trusted platform module, a function for securing a SHVS hardware

**TRP** transmission and reception point

**UDM** unified data management

**UE** user equipment, a 3GPP term normally associated with a mobile handset

**UL** uplink—data that flows from a UE or user in the direction of the network

**UMTS** Universal Mobile Telecommunications System, third generation mobile cellular system

**UP** user plane, the portion of a communication network that transports user data

**UPF** user plane function

**URLLC** Ultra Reliable Low Latency Communications, a 5G concept to improve latency

**USD** United States dollars, not a CSP term, rather more formal way of saying $

**VM** virtual machine, the virtualization or emulation of a computer system inclusive of its OS

**VNF** virtualized network function, a network function that is realized purely in software, to run on a SHVS

**vRAN** the virtualized radio access network, the disaggregation of the radio network

**xApp** RIC microservic

**XDP** eXpress Data Path, an addition to the sockets protocol in the Linux Kernel to speed I/O data transfers

**ZTA** zero trust architecture

# About the Authors

**Larry J. Horner** is a principal engineer and senior solution architect at Intel. His efforts focus on accelerating the transformation of the global communication provider network. His work covers all aspects of the carrier transformation including virtualization from innovation to scale deployment from the enterprise edge premises equipment into the core of the CSP network.

Larry is a trusted advisor to a number of leading global CSP operators and has worked in areas of transformation of the CSP network going back to the earliest days of the mobile network.

He leads training workshops and has over 70 Intel Network Builders courses available and works to accelerate innovative transformations with AT&T, Vz, T-Mobile, BT, Telefonica, NTT, DTAG, Orange, MTS, Singtel, Telstra, and others.

Larry holds degrees in electrical engineering and specialized in information and communication theory beyond the master's level from the University of Akron. He holds two U.S. patents relating to fundamental changes in the communication network transitioning to the all-IP network. He has held various senior level positions in the network equipment provider community in design, operations, test, support, and influencer roles.

He has been a key technical and management contributor to the IP transformation of the communications community for decades. Larry has designed, developed, and supported critical systems for the CSP network that have gone into over 300 networks worldwide. Larry also has direct experience with a U.S.-based Tier 1 carrier. He is a Life Senior Member of the IEEE, member of the board of the local Communication Society, North America Region 5 ComSoc representative, and general cochair for the International IEEE NFV SDN conference. He is a regular speaker at conferences and has briefed the Office of the

President of the United States on telecom security. In his spare time, he enjoys building houses and furniture.

**Kurt Tutschku** is a professor for telecommunication systems at the Blekinge Institute of Technology (BTH). He leads BTH's team on secure and distributed systems.

Prior to BTH, Kurt was a professor at the University of Vienna (endowed by A1 Telekom Austria) and worked with the Network Virtualization Lab of the National Institute for Information and Communication Technology (NICT) in Tokyo, Japan. He received his PhD in computer science in 1999 and his Habilitation degree in 2008. Both degrees were awarded by the University of Würzburg, Germany.

Kurt's research interests are centered around the architectures and mechanism of future generation distributed systems, softwarized networks, and clouds (including NFV). He has carried out or leads multiple funded academic and industry collaborations in this area. Kurt and his team contributed to various Future Internet and network virtualization testbed projects, including GENI (U.S.), Akari (Japan), or G-Lab (Germany). Kurt has published over 90 book, journal, and conference contributions. He has served as one of the general co-chairs of the IEEE Conference on NFV-SDN since 2017.

**Andrea Fumagalli** is a professor of electrical and computer engineering at the University of Texas at Dallas. He holds a PhD in electrical engineering (1992) and a laurea degree in electrical engineering (1987), both from the Politecnico di Torino, Torino, Italy. From 1992 to 1998 he was an assistant professor in the Electronics Engineering Department at the Politecnico di Torino, Italy. He joined UT-Dallas as an associate professor of electrical engineering in August 1997 and was elevated to the rank of professor in 2005.

Dr. Fumagalli has been chosen to lead a subset of an international collaboration to create the computer network architecture of the future—preparing for a time when trillions of devices are expected to be connected to the internet. Dr. Fumagalli's research interests include aspects of wireless, optical, Internet of Things (IoT), cloud networks, and related protocol design and performance evaluation. He has published over 250 technical papers in peer-reviewed refereed journals and conferences.

**ShunmugaPriya Ramanathan** (Priya) is pursuing her doctoral degree in 5G network function virtualization at the University of Texas at Dallas (UTD). She works in the Open Network Advanced Research (OpNeAR) lab under the guidance of Dr. Andrea Fumagalli. Her research focuses on the performance evaluation of various open-source reliability schemas for the virtualized 5G Radio Access Network (RAN) and its corresponding transport network components. She is an active reviewer of several IEEE papers and recently became an IEEE Senior Member. During her research time, she also worked as a graduate intern at Intel for nine months. Her research interests include software engineering, 5G RAN, IoT, fault protection, and restoration with performance evaluation.

Priya received her BE in electronics and communication engineering from Thiagarajar College of Engineering, Madurai, India, in 2004 and her master's in electrical engineering from UTD in 2018. She worked as the technical lead at Honeywell Technology of Solutions. She was a firmware engineer in Home and Building Solutions to build Honeywell's Heating, Ventilation, and Air Conditioning (HVAC) controllers from 2004 to 2012. She also provided L3 support for the HVAC field engineers in the Asia Pacific and North American regions. During her work, she received a U.S. patent (US 8174962 B2) granted in BACnet communication systems, an open-source communication protocol for home and building solutions.

# Index

## Artech House Mobile Communications Library

William Webb, Series Editor

*3G CDMA2000 Wireless System Engineering,* Samuel C. Yang

*3G Multimedia Network Services, Accounting, and User Profiles,* Freddy Ghys, Marcel Mampaey, Michel Smouts, and Arto Vaaraniemi

*5G and Satellite RF and Optical Integration,* Geoff Varrall

*5G-Enabled Industrial IoT Networks,* Amitava Ghosh, Rapeepat Ratasuk, Simone Redana, and Peter Rost

*5G New Radio: Beyond Mobile Broadband,* Amitav Mukherjee

*5G Spectrum and Standards,* Geoff Varrall

*802.11 WLANs and IP Networking: Security, QoS, and Mobility,* Anand R. Prasad and Neeli R. Prasad

*Achieving Interoperability in Critical IT and Communications Systems,* Robert I. Desourdis, Peter J. Rosamilia, Christopher P. Jacobson, James E. Sinclair, and James R. McClure

*Advances in 3G Enhanced Technologies for Wireless Communications,* Jiangzhou Wang and Tung-Sang Ng, editors

*Advances in Mobile Information Systems,* John Walker, editor

*Advances in Mobile Radio Access Networks,* Y. Jay Guo

*Artificial Intelligence in Wireless Communications,* Thomas W. Rondeau and Charles W. Bostian

*Broadband Wireless Access and Local Network: Mobile WiMax and WiFi,* Byeong Gi Lee and Sunghyun Choi

*CDMA for Wireless Personal Communications,* Ramjee Prasad

*CDMA RF System Engineering,* Samuel C. Yang

*CDMA Systems Capacity Engineering,* Kiseon Kim and Insoo Koo

*Cell Planning for Wireless Communications,* Manuel F. Cátedra and Jesús Pérez-Arriaga

*Cellular Communications: Worldwide Market Development,* Garry A. Garrard

*Cellular Mobile Systems Engineering,* Saleh Faruque

*Cognitive Radio Interoperability through Waveform Reconfiguration,* Leszek Lechowicz and Mieczyslaw M. Kokar

*Cognitive Radio Techniques: Spectrum Sensing, Interference Mitigation, and Localization,* Kandeepan Sithamparanathan and Andrea Giorgetti

*The Complete Wireless Communications Professional: A Guide for Engineers and Managers,* William Webb

*EDGE for Mobile Internet,* Emmanuel Seurre, Patrick Savelli, and Pierre-Jean Pietri

*Emerging Public Safety Wireless Communication Systems,* Robert I. Desourdis, Jr., et al.

*From LTE to LTE-Advanced Pro and 5G,* Moe Rahnema and Marcin Dryjanski

*The Future of Wireless Communications,* William Webb

*Geospatial Computing in Mobile Devices,* Ruizhi Chen and Robert Guinness

*GPRS for Mobile Internet,* Emmanuel Seurre, Patrick Savelli, and Pierre-Jean Pietri

*GSM and Personal Communications Handbook,* Siegmund M. Redl, Matthias K. Weber, and Malcolm W. Oliphant

*GSM Networks: Protocols, Terminology, and Implementation,* Gunnar Heine

*GSM System Engineering,* Asha Mehrotra

*Handbook of Land-Mobile Radio System Coverage,* Garry C. Hess

*Handbook of Mobile Radio Networks,* Sami Tabbane

*Handbook of Next-Generation Emergency Services,* Barbara Kemp and Bart Lovett

*High-Speed Wireless ATM and LANs,* Benny Bing

*Implementing Full Duplexing for 5G,* David B. Cruickshank

*In-Band Full-Duplex Wireless Systems Handbook,* Kenneth E. Kolodziej, Editor

*Inside Bluetooth Low Energy, Second Edition,* Naresh Gupta

*Interference Analysis and Reduction for Wireless Systems,* Peter Stavroulakis

*Interference and Resource Management in Heterogeneous Wireless Networks,* Jiandong Li, Min Sheng, Xijun Wang, and Hongguang Sun

*Internet Technologies for Fixed and Mobile Networks,* Toni Janevski

*Introduction to 3G Mobile Communications, Second Edition,* Juha Korhonen

*Introduction to 4G Mobile Communications,* Juha Korhonen

*Introduction to Communication Systems Simulation,* Maurice Schiff

*Introduction to Digital Professional Mobile Radio,* Hans-Peter A. Ketterling

*An Introduction to GSM,* Siegmund M. Redl, Matthias K. Weber, and Malcolm W. Oliphant

*Introduction to Mobile Communications Engineering,* José M. Hernando and F. Pérez-Fontán

*Introduction to OFDM Receiver Design and Simulation,* Y. J. Liu

*An Introduction to Optical Wireless Mobile Communications,* Harald Haas, Mohamed Sufyan Islim, Cheng Chen, and Hanaa Abumarshoud

*Introduction to Radio Propagation for Fixed and Mobile Communications,* John Doble

*Introduction to Wireless Local Loop, Broadband and Narrowband, Systems, Second Edition,* William Webb

*IS-136 TDMA Technology, Economics, and Services,* Lawrence Harte, Adrian Smith, and Charles A. Jacobs

*Location Management and Routing in Mobile Wireless Networks,* Amitava Mukherjee, Somprakash Bandyopadhyay, and Debashis Saha

*LTE Air Interface Protocols*, Mohammad T. Kawser

*Metro Ethernet Services for LTE Backhaul,* Roman Krzanowski

*Mobile Data Communications Systems*, Peter Wong and David Britland

*Mobile IP Technology for M-Business,* Mark Norris

*Mobile Satellite Communications,* Shingo Ohmori, Hiromitsu Wakana, and Seiichiro Kawase

*Mobile Telecommunications Standards: GSM, UMTS, TETRA, and ERMES,* Rudi Bekkers

*Mobile-to-Mobile Wireless Channels,* Alenka Zajić

*Mobile Telecommunications: Standards, Regulation, and Applications,* Rudi Bekkers and Jan Smits

*Multiantenna Digital Radio Transmission,* Massimiliano Martone

*Multiantenna Wireless Communications Systems,* Sergio Barbarossa

*Multi-Gigabit Microwave and Millimeter-Wave Wireless Communications,* Jonathan Wells

*Multiuser Detection in CDMA Mobile Terminals,* Piero Castoldi

*OFDMA for Broadband Wireless Access,* Slawomir Pietrzyk

*Practical Wireless Data Modem Design,* Jonathon Y. C. Cheah

*The Practitioner's Guide to Cellular IoT,* Cameron Kelly Coursey

*Prime Codes with Applications to CDMA Optical and Wireless Networks,* Guu-Chang Yang and Wing C. Kwong

*Quantitative Analysis of Cognitive Radio and Network Performance,* Preston Marshall

*QoS in Integrated 3G Networks,* Robert Lloyd-Evans

*Radio Resource Management for Wireless Networks,* Jens Zander and Seong-Lyun Kim

*Radiowave Propagation and Antennas for Personal Communications, Third Edition,* Kazimierz Siwiak and Yasaman Bahreini

*RDS: The Radio Data System,* Dietmar Kopitz and Bev Marks

*Resource Allocation in Hierarchical Cellular Systems,* Lauro Ortigoza-Guerrero and A. Hamid Aghvami

*RF and Baseband Techniques for Software-Defined Radio,* Peter B. Kenington

*RF and Microwave Circuit Design for Wireless Communications,* Lawrence E. Larson, editor

*Sample Rate Conversion in Software Configurable Radios,* Tim Hentschel

*Signal Processing Applications in CDMA Communications,* Hui Liu

*Signal Processing for RF Circuit Impairment Mitigation,* Xinping Huang, Zhiwen Zhu, and Henry Leung

*Smart Antenna Engineering,* Ahmed El Zooghby

*Software-Defined Radio for Engineers,* Travis F. Collins, Robin Getz, Di Pu, and Alexander M. Wyglinski

*Software Defined Radio for 3G,* Paul Burns

*Spectrum Wars: The Rise of 5G and Beyond,* Jennifer A. Manner

*Spread Spectrum CDMA Systems for Wireless Communications,* Savo G. Glisic and Branka Vucetic

*Technical Foundations of the IoT,* Boris Adryan, Dominik Obermaier, and Paul Fremantle

*Technologies and Systems for Access and Transport Networks,* Jan A. Audestad

*Third-Generation and Wideband HF Radio Communications,* Eric E. Johnson, Eric Koski, William N. Furman, Mark Jorgenson, and John Nieto

*Third Generation Wireless Systems, Volume 1: Post-Shannon Signal Architectures,* George M. Calhoun

*Traffic Analysis and Design of Wireless IP Networks,* Toni Janevski

*Transmission Systems Design Handbook for Wireless Networks,* Harvey Lehpamer

*UMTS and Mobile Computing,* Alexander Joseph Huber and Josef Franz Huber

*Understanding Cellular Radio,* William Webb

*Understanding Digital PCS: The TDMA Standard,* Cameron Kelly Coursey

*Understanding WAP: Wireless Applications, Devices, and Services,* Marcel van der Heijden and Marcus Taylor, Editors

*Universal Wireless Personal Communications,* Ramjee Prasad

*Virtualizing 5G and Beyond 5G Mobile Networks,* Larry J. Horner, Kurt Tutschku, Andrea Fumagalli, and ShunmugaPriya Ramanathan

*WCDMA: Towards IP Mobility and Mobile Internet,* Tero Ojanperä and Ramjee Prasad, editors

*Wi-Fi 6: Protocol and Network,* Susinder R. Gulasekaran and Sundar G. Sankaran

*Wireless Communications in Developing Countries: Cellular and Satellite Systems,* Rachael E. Schwartz

*Wireless Communications Evolution to 3G and Beyond,* Saad Z. Asif

*Wireless Intelligent Networking,* Gerry Christensen, Paul G. Florack and Robert Duncan

*Wireless LAN Standards and Applications,* Asunción Santamaría and Francisco J. López-Hernández, editors

*Wireless Sensor and Ad Hoc Networks Under Diversified Network Scenarios,* Subir Kumar Sarkar

*Wireless Technician's Handbook, Second Edition,* Andrew Miceli

For further information on these and other Artech House titles, including previously considered out-of-print books now available through our In-Print-Forever® (IPF®) program, contact:

Artech House
685 Canton Street
Norwood, MA 02062
Phone: 781-769-9750
Fax: 781-769-6334
e-mail: artech@artechhouse.com

Artech House
16 Sussex Street
London SW1V 4RW UK
Phone: +44 (0)20 7596-8750
Fax: +44 (0)20 7630-0166
e-mail: artech-uk@artechhouse.com

Find us on the World Wide Web at: www.artechhouse.com